Studies in Space Policy

Volume 34

Series Editor

European Space Policy Institute, Vienna, Austria

The use of outer space is of growing strategic and technological relevance. The development of robotic exploration to distant planets and bodies across the solar system, as well as pioneering human space exploration in earth orbit and of the moon, paved the way for ambitious long-term space exploration. Today, space exploration goes far beyond a merely technological endeavour, as its further development will have a tremendous social, cultural and economic impact. Space activities are entering an era in which contributions of the humanities — history, philosophy, anthropology —, the arts, and the social sciences — political science, economics, law — will become crucial for the future of space exploration. Space policy thus will gain in visibility and relevance. The series Studies in Space Policy shall become the European reference compilation edited by the leading institute in the field, the European Space Policy Institute. It will contain both monographs and collections dealing with their subjects in a transdisciplinary way.

The volumes of the series are single-blind peer-reviewed.

More information about this series at https://link.springer.com/bookseries/8167

Annette Froehlich
Editor

Spaceports in Europe

ESPI
European Space Policy Institute

Springer

Editor
Annette Froehlich
European Space Policy Institute (ESPI)
German Aerospace Center (DLR)
Vienna, Austria

ISSN 1868-5307 ISSN 1868-5315 (electronic)
Studies in Space Policy
ISBN 978-3-030-88310-2 ISBN 978-3-030-88311-9 (eBook)
https://doi.org/10.1007/978-3-030-88311-9

This Springer imprint is published by the registered company Springer Nature Switzerland AG
The registered company address is: Gewerbestrasse 11, 6330 Cham, Switzerland

Preface

Europe is seeking to enlarge its launching capacities by building additional spaceports on the European continent. Various national initiatives are envisaged resulting in a "space race" in the field of constructing spaceports and building micro launchers. However, right from the beginning when choosing the launch site (land- or sea-based rocket launches) there are various factors relating to international space law, European regulations, and national rules that must be considered, as spaceports are rarely explicitly addressed in current legal and policy frameworks. While launching sites used to be operated by governments, private commercial initiatives are increasingly entering the field. This paradigm shift must be reflected within regulations relating to various aspects of space liability by enlarging the long-established terms of the United Nations space treaties to accommodate commercial space flights. Questions of permission, supervision, and control require special liability regulations to avoid detrimental consequences stemming from the concept of "launching states" in view of the rise of private driven commercial space activities on a global level. Furthermore, not only do environmental aspects need to be thoroughly examined but also the concept of critical infrastructure requires special attention from a security perspective to anticipate, *inter alia*, cyber-attacks. For these reasons, several European and national regulations may need to be enlarged to apply to the entire space sector, using a harmonized approach that has direct implications for the regulations, programmes, and missions of the European Union and the European Space Agency, bearing in mind that the upcoming spaceports in Europe are an essential asset to substantially boost the European New Space.

Vienna, Austria
August 2021

Dr. Annette Froehlich

Contents

Securing Europe's Spaceports: A Legal Perspective

Jack Wright Nelson

Abstract The European economy is highly reliant on space. Access to space requires access to spaceports. Yet securing this access—that is, spaceport security—is both under-analyzed and under-formalized. In particular, the European Union's critical infrastructure protection directives do not specifically address spaceports. Further, like many modern industries, the entire space sector is now wholly dependent on computer networks. These networks are vulnerable to cyber-attack. A successful attack on a spaceport during fueling, launch or range control operations could cause catastrophic damage to persons and property. Therefore, as spaceports proliferate across Europe, ensuring their physical and cybersecurity is a crucial challenge for the European Union. This chapter provides a legal perspective on this challenge and critiques two current proposals from the European Commission that aim to incorporate certain space infrastructures within existing frameworks.

1 Introduction

From Taranto-Grottaglie to Kiruna to the Azores, spaceports are spreading across the European Union ("EU"). Soon, the European Space Agency's ("ESA") christening of the Centre Spatial Guyanais ("CSG") as 'Europe's Spaceport' will sound anachronistic. By the end of the decade, a satellite operator, proudly announcing the launch of their new fleet from "Europe's Spaceport" may prompt the reply: 'Which one?'.

These new spaceports face an uneven regulatory topology. Traditionally, spaceports have been built, owned and operated by governments. This arrangement sidestepped the need for dedicated regulatory regimes. Instead, governments have often relied on bespoke, internal policies to govern their spaceports. An example of this is the CSG itself. Before 2010, this spaceport operated within the infamous "French paradox"—that is, the situation where "more and more (private) space activities were taking place from the French territory [of Guyana] without being subject

J. W. Nelson (✉)
Faculty of Law, National University of Singapore, Singapore, Singapore

A. Froehlich (ed.), *Spaceports in Europe*, Studies in Space Policy 34,
https://doi.org/10.1007/978-3-030-88311-9_1

to specific regulation."[1] The lack of spaceport-specific regulation is not surprising. D'Angelo (a former official at the United Nations Office of Outer Space Affairs) lamented in 1994 that while "commercial activities in space have greatly expanded, the applicable laws lag far behind."[2] This remains true today, both within Europe and further afield. Spaceport regulation, as Seedhouse summarizes, "is very much a work in progress because there are so many national, regional and supranational legal mechanisms that must be considered."[3] Given that the global economy is now highly dependent on space services, it seems that what was an initially a French paradox is increasingly a global one.

One consequence of this uncertain regulatory environment is that spaceport security is under-analyzed and under-formalized, particularly from a legal perspective. Yet spaceport security is paramount: disruption to a spaceport could have cascading effects on everything from telecommunications to ridesharing applications and the global financial system.[4] The pivotal role of spaceports means that they increasingly meet the standard criteria to be considered as "critical infrastructure" ("CI")—that is, assets that are essential to ensure security, enable economic activity and protect public health and safety.[5] Further, the space sector exhibits a very high degree of digital intensity. This makes spaceports likely targets for cyber-attacks.[6] In the specific case of spaceports, Zatti (former head of the ESA Security Office) describes the unauthorized operation of ground facilities as "one of the most feared threats among the flight control team members and possibly one of the easiest ones to achieve."[7] He continues: "[i]f someone managed to insert a wrong command into the launch-automated sequent to inject a failure on any of the critical activities, like the separation sequence, or the burns needed, then it could cause the loss of the launcher and of the spacecraft carried inside."[8]

Indeed, any attack—cyber or otherwise—during launch operations could cause catastrophic damage, on the ground, in airspace and in outer space.

The two primary EU directives relating to CI—the European critical infrastructure directive ("ECI Directive")[9] and network and information security directive ("NIS

---

[1] Michael Gerhard and Isabelle Reutzel, 'Law Related to Space Transportation and Spaceports' in Ram S Jakhu and Paul S Dempsey (eds), *Routledge Handbook of Space Law* (Routledge 2016) 274.

[2] George V D'Angelo, *Aerospace Business Law* (Quorum Books 1994) 105.

[3] Erik Seedhouse, *Spaceports Around the World, A Global Growth Industry* (Springer 2016) 34.

[4] For instance, if a telecommunications satellite requires replacement, but an attack on a spaceport disrupts launch operations, then the replacement may not be orbited in time to avoid outages.

[5] Alexandru Georgescu and others, *Critical Space Infrastructures: Risk, Resilience and Complexity*, vol 36 (Springer 2019) 11.

[6] Jacob G Oakley, *Cybersecurity for Space: Protecting the Final Frontier* (Apress 2020).

[7] Stefano Zatti, 'Space and Cyber Threats' in Kai-Uwe Schrogl (ed), *Handbook of Space Security: Policies, Applications and Programs* (2nd edn, Springer 2020) 253.

[8] ibid.

[9] Council Directive 2008/114/EC of 8 December 2008 on the Identification and Designation of European Critical Infrastructures and the assessment of the Need to Improve their Protection.

Directive")[10]—do not directly apply to spaceports or the broader space industry. Similarly, the primary space sector-specific EU instrument—the regulation establishing the Union Space Programme and the European Union Agency for the Space Programme ("Space Programme Regulations")[11]—does not target spaceports. But changes are underway. Proposed revisions to the ECI Directive and the NIS Directive—which I will respectively refer to as "ECI 2.0"[12] and "NIS 2.0"[13]—specify "[s]pace" as a relevant sector. Under both ECI 2.0 and NIS 2.0, this broad term covers "[o]perators of ground-based infrastructure, owned, managed and operated by Member States or by private parties, that support the provision of space-based services."[14] While spaceport operators would likely fall within this scope, my argument is that the ECI 2.0 and NIS 2.0 proposals are unlikely to provide the optimal legal framework to secure Europe's spaceports. More specifically, the broad approach proposed for ECI 2.0 and NIS 2.0 simply specifies "space" as a standalone sector without specifying any subsectors. This approach fails to recognize that spaceports have unique risk profiles and dependencies. A more granular approach is warranted. To this end, I argue for the inclusion of spaceports as a subsector within the existing "[t]ransport" sector to which both the ECI Directive and the NIS Directive currently apply.

I have structured this chapter as follows. In Sect. 2, I provide an overview of spaceports and their regulation. I then outline the ECI and NIS Directives in Sect. 3, before assessing the adequacy of ECI 2.0 and NIS 2.0 vis-à-vis spaceports in Sect. 4. Section 5 concludes.

2 Background and Regulatory Environment

For the benefit of readers who are not industry specialists, this section provides a brief overview of spaceports and their regulation.

---

[10] Directive (EU) 2016/1148 of the European Parliament and of the Council of 6 July 2016 Concerning Measures for a High Common Level of Security of Network and Information Systems across the Union.

[11] Regulation (EU) 2021/696 of the European Parliament and of the Council of 28 April 2021 establishing the Union Space Programme and the European Union Agency for the Space Programme and repealing Regulations (EU) No 912/2010, (EU) No 1285/2013 and (EU) No 377/2014 and Decision No 541/2014/EU.

[12] Proposal for a Directive of the European Parliament and of the Council on the Resilience of Critical Entities (COM(2020) 829 final).

[13] Proposal for a Directive of the European Parliament and of the Council on Measures for a High Common Level of Cybersecurity Across the Union, Repealing Directive 2016/1148 (COM(2020) 823 final).

[14] Annex to the Proposal for a Directive of the European Parliament and of the Council on the Resilience of Critical Entities (COM(2020) 829 final) 7; Annexes to the Proposal for a Directive of the European Parliament and of the Council on Measures for a High Common Level of Cybersecurity Across the Union, Repealing Directive 2016/1148 (COM(2020) 823 final) 8.

2.1 Spaceports

At a high level, a spaceport can be defined as:

> a transportation hub where all the elements for a spaceflight come together. First, components arrive by air, rail, water or by road. Next, rocket elements are integrated to create the launch vehicle. Propellants are then transferred to the launch vehicle, the crew boards and the vehicle launches.[15]

While this refers to traditional human spaceflight, various launches take place from modern spaceports. They range from small satellite launches to suborbital space tourism and interplanetary missions. By way of example, three different rockets currently launch from the CSG, serving a range of governmental, scientific and commercial purposes.[16]

Not all spaceports provide traditional, land-based rocket launches. Some spaceports support horizontal launches and returns.[17] Floating spaceports are a further alternative. A prime example of this is the sea segment of the Luigi Broglio Space Center, a launch platform located off the Kenyan coast that is managed by the Agenzia Spaziale Italiana ("ASI").[18] While floating spaceports remain relatively rare, this may change over time. As Pozdnakova notes, "[t]he increasing global demand" for space access "may partly be met by floating spaceports, ie vessels or platforms located at sea from which satellites can be launched into orbit."[19]

The range of ancillary activities undertaken at spaceports is expanding. In particular, space tourists are likely to need places to train and, as Seedhouse wryly remarks, "a suitably high-end resort [in which] to stay."[20] Spaceports will need to be more than just a launchpad: ultimately, "[a]n environment similar to a commercial airport, or a cruise ship terminal, will be needed."[21] These developments mean that the traditional practices of securing spaceports by co-locating them with military facilities, or by deploying military personnel, are unlikely to continue.[22] This further emphasizes the importance of addressing spaceport security in its own right and within its own paradigm—a theme I will return to later in this chapter.

---

[15] Seedhouse (n 5) 6.

[16] Ariane 5, Soyuz and Vega. See generally 'The Spaceport' (*Arianespace*) < https://www.arianespace.com/spaceport-facility/ > accessed 28 July 2021.

[17] This involves spacecraft launching into space from carrier aircraft. The carrier aircraft take off and land from runways.

[18] See generally 'Luigi Broglio Space Center' (*Agenzia Spaziale Italiana*) < https://www.asi.it/en/the-agency/the-space-centers/luigi-broglio-space-center/ > accessed 30 July 2021.

[19] Alla Pozdnakova, 'Oceans as Spaceports: State Jurisdiction and Responsibility for Space Launch Projects at Sea' (2020) 26 Journal of International Maritime Law 267, 267.

[20] Seedhouse (n 5) 2.

[21] ibid 3.

[22] For example, the well-known John F. Kennedy Space Centre is adjacent to Cape Canaveral Space Force Station and the CSG is secured by personnel from France's Légion étrangère.

2.2 *Spaceport Regulation*

The international treaties that govern space activities do not address spaceports specifically. This omission is not surprising: the last of these treaties was concluded in 1979,[23] when there were few spaceports (and all were in government hands). Nonetheless, there are specific provisions of these treaties that are relevant to states housing spaceports—in particular, liability under the Outer Space Treaty[24] ("OST") and the Liability Convention[25] ("LIAB").

First, OST article VII provides that:

> [e]ach State Party to the Treaty that launches or procures the launching of an object into outer space, including the moon and other celestial bodies, and each State Party from whose territory or facility an object is launched, is internationally liable for damage to another State Party to the Treaty or to its natural or juridical persons by such object or its component parts on the Earth, in air or in outer space, including the moon and other celestial bodies.

Second, LIAB article II provides that:

> [a] launching State shall be absolutely liable to pay compensation for damage caused by its space object on the surface of the [E]arth or to aircraft in flight.

Under LIAB, a "launching State" includes "[a] State from whose territory or facility a space object is launched."[26] Further, article V imposes joint and several liability "whenever two or more States jointly launch a space object" and deems a State "from whose territory or facility a space object is launched" to be "a participant in a joint launching". Accordingly, a state that has a spaceport on its territory faces extensive potential liability under international law. A state with a "facility" from which launches occur—such as a floating spaceport—faces the same potential liability.

To meet this potential liability, states that regulate spaceports have typically imposed comprehensive insurance and indemnification requirements on spaceport operators. Further, some states have regulated to ensure the protection of people, property and the environment. In addition, the licensing authority usually reviews the operator's technical and financial adequacy.[27]

Within Member States, there is generally little space-specific regulation. Austria is a notable exception. The Bundesgesetz über die Genehmigung von Weltraumaktivitäten und die Einrichtung eines Weltraumregisters ("Austrian Outer Space Act")[28]

---

[23] Agreement Governing the Activities of States on the Moon and Other Celestial Bodies (adopted 5 December 1979, entered into force 11 July 1984) 1363 UNTS 3.

[24] Treaty on Principles Governing the Activities of States in the Exploration and Use of Outer Space, including the Moon and other Celestial Bodies (adopted 27 January 1967, entered into force 10 October 1967) 610 UNTS 205.

[25] Convention on International Liability for Damage Caused by Space Objects (adopted 29 March 1972, entered into force 1 September 1972) 961 UNTS 13,810.

[26] LIAB, art 1.

[27] Gerhard and Reutzel (n 3) 285.

[28] BGBl. I Nr. 132/2011.

specifies that it applies to the "Betrieb einer Anlage zum Start von Weltraumgegenständen,"[29] and that such operation requires authorization from the Austrian Minister for Transport, Innovation and Technology.[30] Regulations made under the Austrian Outer Space Act[31] further specify that the application for this authorization must be accompanied by.

> ein Zertifikat über eine erfolgte Sicherheitsüberprüfung im Sinne des Sicherheitspolizeigesetzes, BGBl. Nr. 566/1991, in der geltenden Fassung, oder Verlässlichkeitsprüfung im Sinne des Militärbefugnisgesetzes, BGBl. I Nr. 86/2000, in der geltenden Fassung, des Betreibers oder, soweit es sich um eine juristische Person handelt, des für die Weltraumaktivität verantwortlichen Vertreters.[32]

Accordingly, Austria's space legislation both specifically applies to spaceports and mandates a security review prior to authorizing such spaceports. This spaceport and security specific coverage is relatively rare. For example, the relevant French law governing space activities[33] does not specifically regulate spaceports. Rather, it principally governs "[o]pération[s] spatiale[s]" which are broadly defined as:

> toute activité consistant à lancer ou tenter de lancer un objet dans l'espace extra-atmosphérique ou à assurer la maîtrise d'un objet spatial pendant son séjour dans l'espace extra-atmosphérique, y compris la Lune et les autres corps célestes, ainsi que, le cas échéant, lors de son retour sur Terre.[34]

Further, the specific law applying to the CSG simply grants the president of France's Centre national d'études spatiales ("CNES") "la police spéciale de l'exploitation des installations du [CSG] dans un périmètre délimité par l'autorité administrative compétente."[35] While this arrangement has ensured the security of the CSG, it is not practicable nor appropriate for future, privately-owned and operated spaceports.

[29] Austrian Outer Space Act, § 2(1). In English, "the operation of a launch facility".

[30] Austrian Outer Space Act, § 3.

[31] *Verordnung der Bundesministerin/des Bundesministers für Verkehr, Innovation und Technologie zur Durchführung des Bundesgesetzes über die Genehmigung von Weltraumaktivitäten und die Einrichtung eines Weltraumregisters*, BGBl. II Nr. 36/2015 ("Austrian Outer Space Regulations").

[32] Austrian Outer Space Regulations, § 2. (1). In English, "a certificate about a security review undertaken in accordance with the Federal Security Police Act, BGBl. No. 566/1991, as amended, or of a reliability review in accordance with the Federal Military Authority Act, BGBl. I No. 86/2000, as amended, of the operator, or, in case of a juridical person, of the representative responsible for the space activity".

[33] Loi n° 2008–518 du 3 juin 2008 relative aux opérations spatiales ("French Space Operations Act").

[34] French Space Operations Act, art 1(3). In English, "any activity of launching or attempting to launch an object into outer space or of exercising control over a space object while it is in outer space, including the Moon and other celestial bodies, as well as, if applicable, during its return to Earth."

[35] Code de la recherche, art L331-6. In English, "special police powers for the operations of the installations of the [CSG] within a perimeter delimited by the competent administrative authority."

Looking beyond the EU, Australia,[36] New Zealand,[37] the United Kingdom,[38] and the United States[39] all have spaceport-specific regulatory regimes. But major space-capable states, such as India and China, do not. Surveying this uneven regulatory topography, Gerhard and Reutzel conclude that the critical challenge is "achieving an internationally harmonized set of rules" for spaceports given that the "lack of a detailed and uniform international regulatory system, including technical standards" means that "there is no guarantee of uniformity."[40]

The fragmented regulatory landscape is a particular issue when it comes to securing Europe's spaceports. Digitization and market liberalization mean that spaceports and other CIs are part of extensive physical and digital networks that spread across Europe (and beyond). Public and private sector reliance on these networks is well-known and pervasive. However, the security of a network is only as strong as its weakest link. Accordingly, a common, uniform level of protection is not just beneficial; it is necessary to protect the network. This fundamental insight into the cross-border nature of Europe's CI is the premise for the ECI Directive and the NIS Directive. I address both directives in the following section.

3 The ECI and NIS Directives

A primary mechanism to increase legal uniformity across the EU is the negotiation and agreement of directives.[41] The ECI Directive and the NIS Directive both aim to promote CI protection by setting out baseline protection measures. The former does this directly, while the latter targets cybersecurity. However, both are limited to specific sectors. At present, neither includes space as a specified sector.

---

[36] Space (Launch and Returns) Act 2018, s 11 (referring to a "launch facility").

[37] Outer Space and High-Altitude Activities Act 2017, s 38 (referring to a "launch facility").

[38] Space Industry Act 2018, s 3(1)(b) (referring to "spaceports").

[39] 51 United States Code § 50,904(a) (referring to "launch sites").

[40] Gerhard and Reutzel (n 3) 287.

[41] Directives are EU legislative acts that set out specific goals that all Member States must achieve. It remains within the purview of the Member States to devise their own laws on how to reach these goals. See generally 'Regulations, Directives and Other Acts' (*European Union*, 16 June 2016) < https://europa.eu/european-union/law/legal-acts_en > accessed 28 July 2021.

3.1 ECI Directive

On 17 November 2005, the European Commission adopted a green paper[42] on a European programme for critical infrastructure protection ("Green Paper").[43] The Green Paper suggested that a common framework for CI protection would strengthen CI in the EU.[44] The Green Paper identified 11 sectors to form part of this common framework. One of these sectors was "Space and Research".[45] The initial proposal from the European Commission[46] was, in line with the Green Paper, comprehensive.[47] However, following "controversial discussions in [the] Council [of the European Union]",[48] the European Commission elected to focus on the transport and energy sectors.[49] As a result, the ECI Directive "establishes a procedure for the identification and designation of European critical infrastructures ("ECIs")"[50] but is only implemented for "the energy and transport sectors."[51] Strictly speaking, this does not mean that a spaceport cannot meet the definition of ECI. Indeed, the definition is quite broad:

> [ECI] ... means critical infrastructure located in Member States the disruption or destruction of which would have a significant impact on at least two Member States. The significance of the impact shall be assessed in terms of cross-cutting criteria. This includes effects resulting from cross-sector dependencies on other types of infrastructure.[52]

Critical infrastructure is in turn defined to mean:

> an asset, system or part thereof located in Member States which is essential for the maintenance of vital societal functions, health, safety, security, economic or social well-being of people, and the disruption or destruction of which would have a significant impact in a Member State as a result of the failure to maintain those functions.[53]

The CSG (for example) would seem to fall within these definitions. However, the CSG cannot be designated as an ECI because it does not come under the energy

[42] Green papers are European Commission documents that aim to stimulate discussion on particular topics across the EU. See generally 'Green Paper' (*European Union*) < https://eur-lex.europa.eu/summary/glossary/green_paper.html > accessed 28 July 2021.

[43] 'Green Paper on a European Programme for Critical Infrastructure Protection' (Commission of the European Communities 2005) COM(2005) 576 Final.

[44] ibid 5.

[45] This sector was further and unhelpfully separated into two products: "Space" and "Research".

[46] Irmgard Anglmayer, 'European Critical Infrastructure—Revision of Directive 2008/114/EC' (European Parliamentary Research Service 2021) Briefing PE 662.604 4.

[47] ibid.

[48] ibid.

[49] 'Communication from the Commission on a European Programme for Critical Infrastructure Protection' (Commission of the European Communities 2006) COM(2006) 786 Final, 10.

[50] ECI Directive, art 1.

[51] ECI Directive, art 3(3). Anglmayer notes that the ECI Directive's "design allows for extension to other sectors" – at least "theoretically". See: Anglmayer (n 47) 1.

[52] ECI Directive, art 2(b).

[53] ECI Directive, art 2(a).

or transport sectors.[54] The practical impact of designation of an ECI is that, once identified and designated, the relevant Member State must assess that designated ECI's "operator security plan" ("OSP"), or "equivalent measures" to an OSP, in accordance with Annex II to the ECI Directive.[55] If the designated ECI lacks an OSP or equivalent, the relevant Member State "shall ensure by any measures deemed appropriate, that the OSP or equivalent is prepared."[56] In addition, a "Security Liaison Officer" ("SLO") or equivalent needs to be identified for each designated ECI, to "function as a point of contact for security related issues between the owner/operator of the ECI and the relevant Member State authority."[57] Similar to the requirements regarding OSPs, Member States must assess whether "each designated ECI located on its territory" has an SLO or equivalent.[58] If the designated ECI lacks an SLO or equivalent, the relevant Member State "shall ensure by any measures deemed appropriate" that one is designated.[59]

Together with various reporting[60] and consultation[61] requirements, the OSP and SLO requirements represent the common level of protection that must be afforded to designated ECIs across the EU. Of course, Member States are free to adopt requirements that exceed or complement the ECI Directive. Indeed, they may be required to do so under other, sector-specific EU legislation or treaties (for example, concerning nuclear energy).[62]

Interestingly, the Space Programme Regulation provides in Article 34(6)(a) that:

> [t]o ensure the protection of the ground infrastructures which form an integral part of the Programme and which are located on their territory the Member States shall:
>
> (a) an entity provides a service which is essential for the maintenance of critical and/or economic activities;
>
> (i) the protection of European critical infrastructures within the meaning of [the ECI Directive]; and
>
> (ii) the protection of their own national critical infrastructures.

However, the "Programme" is limited to specified components of the EU's space program (such as the Galileo global navigation satellite system).[63] Accordingly, the current situation is as follows: while proposed in the Green Paper for inclusion, the broader space sector is currently outside of the ECI Directive—despite the limited

---

[54] Rather, in the odd phrasing of the ECI Directive, it is an "non-ECI". See ECI Directive, Annex III.

[55] ECI Directive, art 5(2).

[56] ECI Directive, art 5(3).

[57] ECI Directive, art 6(1).

[58] ECI Directive, art 6(2).

[59] ECI Directive, art 6(3).

[60] ECI Directive, art 7.

[61] ECI Directive, art 4.

[62] ECI Directive, preamble (12).

[63] Space Programme Regulations, art 3.

extension of the ECI Directive by the Space Programme Regulation. The situation is similar for the NIS Directive, to which I now turn.

3.2 NIS Directive

The NIS Directive "lays down measures with a view to achieving a high common level of security of network and information systems within the Union so as to improve the functioning of the internal market."[64] These measures include, among other things, that Member States identify "the operators of essential services with an establishment on their territory."[65] This list of operators of essential services is to be updated "on a regular basis, and at least every two years."[66] Member States are also required to ensure that operators of essential services comply with certain security and incident notification requirements.[67]

Who is an "operator of an essential service"? Despite its broad scope, this term is defined restrictively as "a public or private entity of a type referred to in Annex II, which meets the criteria laid down in Article 5(2)." Annex II to the NIS Directive sets out seven different sectors and various subsectors and relevant entity types. For example, the "[t]ransport" sector has four subsectors (air, rail, water and road), each of which has at least two entity types. The "operator of an essential service" definition also refers to the Article 5(2) criteria, which are as follows:

> (a) an entity provides a service which is essential for the maintenance of critical and/or economic activities;
>
> (b) the provision of that service depends on network and information systems; and
>
> (c) an incident would have significant disruptive effects on the provision of that service.

Unlike the ECI Directive, there is no requirement that a disruption to the service would affect multiple Member States. In any event, a European spaceport operator would likely meet these criteria. Nonetheless, such an operator is not an operator of an essential service under the NIS Directive because Annex II omits spaceports and the broader space sector. This situation will change following the implementation of NIS 2.0. I discuss NIS 2.0 and ECI 2.0 in the following section.

---

[64] NIS directive, art 1(1).

[65] NIS Directive, art 5(1).

[66] NIS Directive, art 5(5).

[67] NIS Directive, art 14.

4 ECI 2.0 and NIS 2.0

In July 2019, the European Commission concluded that the ECI Directive is only partially effective and relevant.[68] This negative evaluation followed a previous similar assessment during the 2017 comprehensive review of EU security policy.[69] The ECI Directive's shortcomings can be broadly attributed to changes in the "security context in which [CIs] operate ... since the time the [ECI Directive] entered into force."[70] Similarly, digital intensity has increased significantly since 2016, when the NIS Directive came into effect. Digitalization has accelerated throughout the COVID-19 pandemic. In recognition of this, in late 2020, the European Commission adopted NIS 2.0—in parallel with ECI 2.0—as.

> part of a package of measures to improve further the resilience and incident response capacities of public and private entities, competent authorities and the Union as a whole in the field of cybersecurity and [CI] protection.[71]

Amongst other initiatives, ECI 2.0 forgoes the asset-based approach of the ECI Directive. Instead, ECI 2.0 adopts the NIS Directive's systems-focused approach and specifies particular entity types to which the directive will apply. However, both ECI 2.0 and NIS 2.0 will still only apply within specified sectors. To this end, ECI 2.0 and NIS 2.0 both propose to expand their applicable sectors to include "[w]aste water", "[p]ublic administration" and—most importantly for this chapter—"[s]pace".

These new sectors do not have any subsectors. And just one entity type is specified for the space sector, as follows:

> [o]perators of ground-based infrastructure, owned, managed and operated by Member States or by private parties, that support the provision of space-based services, excluding providers of public electronic communications networks referred to in point (8) of Article 2 of Directive (EU) 2018/1972.[72]

This broad language likely covers spaceport operators because spaceports are "ground-based infrastructure ... that support the provision of space-based services" and do not generally fall into the "electronic communications networks" exclusion. Further, the reference to "Member States or by private parties" will likely carve out any space infrastructure that is "owned, managed and operated" by or on behalf

---

[68] Anglmayer (n 47) 1.

[69] 'Comprehensive Assessment of EU Security Policy' (European Commission 2017) SWD (2017) 278 Final 25. See also Anglmayer (n 47) 5.

[70] Anglmayer (n 47) 1.

[71] NIS 2.0 1.

[72] NIS 2.0 Annexes 8. Point (8) of art 2 of the referenced directive states that "'public electronic communications network' means an electronic communications network used wholly or mainly for the provision of publicly available electronic communications services which support the transfer of information between network termination points." Inexplicably, ECI 2.0 replicates this language but replaces "referred to in point (8)" with "within the meaning of point (8)". This appears to be a distinction without a difference. As such, the change in wording is unnecessary and should be reversed.

of the Union itself.[73] At present, this carve-out is unlikely to affect any current or prospective European spaceport.[74]

ECI 2.0 and NIS 2.0's express reference to "operators of ground-based infrastructure" excludes entities that solely operate space-based infrastructure. Rather, ECI 2.0 and NIS 2.0 target either ground-based uplinks or downlinks to such space-based infrastructure. This terrestrial focus avoids the problematic jurisdictional questions that can arise for space-based assets. However, the reference to "ground-based infrastructure" may also exclude a floating seaport, such as the ASI's Luigi Broglio Space Center. This exclusion appears to be an oversight, given that the operators of terrestrial and floating spaceports have a similar cybersecurity profile. Nonetheless, the practical impact of this exclusion is negligible—floating seaports typically rely on some shore-based infrastructure (such as telemetry, tracking and control facilities) to which NIS 2.0 could apply.[75]

Overall, the inclusion of the space sector in ECI 2.0 and NIS 2.0 is a welcome development. In particular, ECI 2.0 corrects the ECI Directive's deviation from the Green Paper. For its part, NIS 2.0 recognizes an important point aptly made by Zatti: that "the cybersecurity of space missions is a matter of competitiveness for the European space industry."[76] Leaving aside the precise legal basis, instrument choice, and other factors that feed into the EU legislative process, the extension of the ECI Directive and the NIS Directive to the space sector is further justified by reference to three key considerations.

First, the uneven regulatory topology for space activities has resulted in, as Georgescu notes, "gaps" that stem "from the organic development of the space security governance framework."[77] The lack of baseline standards, coupled with the heavy reliance on space-related CIs, means extending CI protection initiatives to space-related CI is "a natural outgrowth".[78]

Second, the space sector has advanced significantly since the ECI Directive entered into force in 2009. To this end, a 2019 evaluative study of the ECI Directive ("Evaluative Study") found that:

---

[73] The extension of the ECI Directive per the Space Programme Regulations would likely fill this gap with respect to Union infrastructure.

[74] At the CSG, for example, the EU is an important customer. But ultimately, it is one among many. The CSG is not owned, managed or operated by or on behalf of the EU. Rather, the CSG is operated by CNES, while the ESA owns the CSG's launch infrastructure for Ariane 5, Soyuz and Vega. See generally: 'Who Does What?' (*Arianespace*) < https://www.arianespace.com/spaceport-facility/who-does-what/ > accessed 29 July 2021.

[75] This is certainly the case with the ASI's Luigi Broglio Space Center, which has both a sea segment and a land segment: see 'Luigi Broglio Space Center' (n 19).

[76] Zatti (n 8) 262.

[77] Alexandru Georgescu, 'Critical Space Infrastructures' in Kai-Uwe Schrogl (ed), *Handbook of Space Security: Policies, Applications and Programs* (2nd edn, Springer 2020) 242.

[78] ibid.

> [t]en years after its entry into force, the [ECI] Directive appears today to have partial to limited relevance, notably in view of recent technological, economic, social, policy/political and environmental developments and current challenges.[79]

The Evaluative Study states explicitly that an area for urgent attention is the space sector,[80] on the grounds that "[i]nfrastructure across Europe is increasing managed electronically and through space-based systems."[81] The NIS Directive is more recent, having entered into force in 2016. Nonetheless, advancements in the capabilities of commercial space enterprises have been particularly noteworthy even since then.[82]

Third, the economies of scale that prevail for space activities mean that the space sector exhibits a strong pan-European orientation.[83] Overall, it is more likely than not that an incident at a European space-related CI would affect two or more Member States.[84] This characteristic warrants a common European standard.

I welcome the extension of the ECI Directive and the NIS Directive to the space sector. However, there are at least two problems with the broad approach taken—that is, specifying "space" in ECI 2.0 and NIS 2.0 as a standalone sector, without specifying any subsectors—when it comes to securing Europe's spaceports. I will illustrate these problems by contrasting spaceports with an equally important, but significantly different piece of ground-based space infrastructure: a satellite dish. This comparison is appropriate because ECI 2.0 and NIS 2.0 will cover spaceports and satellite dishes within a single sector and single entity type.

First, spaceport operators face a risk profile that is distinct from that faced by satellite dish operators. Spaceports have exponentially more attack vectors—and likely a larger impact if any of those attacks were successful. Simply put, the means of attacking (whether physically or otherwise) these two pieces of infrastructure differ significantly.

Second, there are different dependencies between spaceports and satellite dishes. Spaceports have three critical dependencies. First, physical dependencies with the transport nodes that bring passengers, vehicles, payloads and fuel to the spaceport. Second, geographic dependencies with airports and maritime ports (with which they share airspace and the sea). Third, cyber dependencies due to extensive use of information and communication technologies. By contrast, satellite dishes do not generally have extensive physical or geographic dependencies.[85] Further, the nature and

---

[79] Ernst & Young and RAND Europe (n 2) 67.

[80] ibid 83.

[81] Ernst & Young and RAND Europe, 'Evaluation Study of Council Directive 2008/114 on the Identification and Designation of European Critical Infrastructures and the Assessment of the Need to Improve Their Protection' (European Commission 2019) 31.

[82] See, for example, the first crewed, commercial orbital spaceflight – SpaceX Demo-2 – in 2020.

[83] ESA and the various European multinational aerospace enterprises, as well as the recently established European Union Agency for the Space Programme, illustrate this orientation.

[84] Indeed, a security incident at the CSG would affect multiple European states (and more globally, given the widespread usage of CSG for commercial launches).

[85] Although some satellite dishes do require reduced radiofrequency activity.

complexity of cyber dependencies differs drastically between spaceports and satellite dishes.

These differences in risk profile and dependencies justify adopting a more granular approach. In particular, in recognition of the increasing integration of spaceports with other transport forms, I propose that the transport sector under ECI 2.0 and NIS 2.0 include space as a subsector and spaceport operators as a specified entity type.

At present, the ECI Directive and NIS Directive both apply to the transport sector. Under ECI 2.0 and NIS 2.0, transport remains a specified sector divided into four subsectors: air, rail, water and road. Each of these subsectors has various entity types specified. For example, the air subsector under ECI 2.0 and NIS 2.0 specifies entity types such as "[a]ir carriers", [a]irport managing bodies" and "entities operating ancillary installations contained within airports".

To better secure Europe's spaceports, I suggest that ECI 2.0 and NIS 2.0 add a fifth subsector, space, under the transport sector. This space subsector could then specify the applicable entity types in line with the air subsector—for example, "spaceport managing bodies" and "entities operating ancillary installations within spaceports". The current proposal to include a broader space sector, without subsectors, could be retained in its current form but with an exclusion for spaceports covered under the transport sector.

This more granular approach would differentiate spaceports from other ground-based space infrastructures, thereby recognizing the different risk profiles and dependencies identified above. Further, locating spaceports within the transport sector would acknowledge the existing interconnections between the various forms of transport. Indeed, these interconnections will accelerate over time as spaceports proliferate and further integrate themselves within broader transport networks. Treating spaceport security akin to airport security will facilitate this integration and provide a wealth of precedents and procedures for securing Europe's spaceports. More specifically, including space as a transport subsector will invite Member States to leverage their existing experience protecting air, sea and land transport CIs.

5 Final Remarks

The various EU institutions are currently negotiating ECI 2.0 and NIS 2.0. Once the proposals are finalized and adopted, Member States will have 18 months to transpose the revised directives into national law.[86] Amongst these developments, this chapter has provided a legal perspective on the protection of Europe's spaceports. More specifically, I have proposed that ECI 2.0 and NIS 2.0 adopt a more granular approach that differentiates spaceports from other forms of ground-based space infrastructure. I have argued that such an approach is necessary because spaceports face a unique risk profile and have extensive dependencies. In my view, the granular approach would provide a more effective baseline for protecting spaceports within and across

[86] ECI 2.0, art 24; NCI 2.0, art 38.

the Member States. By identifying spaceports within the broader transport sector, this approach would also recognize the increasing integration of spaceports within Europe's traditional transportation networks.

But spaceport security is about more than just law. A broad spectrum of input is needed to protect spaceports (and other CIs). As the Green Paper remarks:

> [t]he effective protection of [CI] requires communication, coordination, and cooperation nationally and at [the] EU level among all interested parties - the owners and operators of infrastructure, regulators, professional bodies and industry associations in cooperation with all levels of government, and the public.[87]

To this end, specifying spaceports as a transportation subsector could raise awareness of spaceports and their security, generate best practice and increase coordination across the Member States. It invites productive comparisons with existing CI protection regimes in the transport sector. More broadly, differentiating spaceports from other ground-based space infrastructures could precipitate a timely discussion across Europe regarding spaceport security. Such a discourse will be necessary for Europe to maintain its independent space access (and its prime position in the global space industry): access to space, after all, requires access to spaceports.

Jack Wright Nelson is a Research Associate at the National University of Singapore Faculty of Law. As a member of the Faculty's Centre for Banking & Finance Law, his research focuses on the regulation and financing of the satellite industry. He is grateful to the Centre for its continued support.

[87] 'Green Paper on a European Programme for Critical Infrastructure Protection' (n 44) 2.

Europe's Spaceports/Launchers and Agreements Referred to in Article V(2) of the Liability Convention

Tugrul Cakir

Abstract According to Article V(2) of the Liability Convention, States may adopt internal agreements to share their high level of liability in relation to joint launches. The conclusion of such agreements would be of utmost importance in the European context given the multiplication of spaceports and micro launcher projects. This contribution addresses how bilateral agreements containing provisions regulating the apportionment of liability have played a fundamental role in space risk management in the European framework before underlining the significance of these agreements within the functioning of the space liability régime. Finally, it analyzes the possible impact of bilateral agreements on the elaboration of a coordinated approach in the new European context.

1 Introduction

To respond to market trends in the field of micro satellites, established launch operators have offered big rocket rideshares. Another response to the miniaturization of satellites has been the development of micro launchers offering more flexibility to customers. Following this tendency, more and more privately led European entities are getting involved in the launch of micro satellites. Simultaneously, the number of commercial spaceport projects is increasing in the European framework.[1] Contrary to spaceports established and operated by States, commercial spaceports are mainly owned and operated by private initiatives.[2] This new European context could lead to four different scenarios.

[1]See further on commercial spaceport projects in Europe ESPI, "Yearbook 2020—Space policies, issues and trends" (June 2021) 65–69.

[2]This involves a paradigm shift in space access. See further ESPI Briefs No. 15, "About the Enthusiastic Multiplication of Spaceports Worldwide" (July 2017).

T. Cakir (✉)
Faculty of Law, Ankara Yildirim Beyazit University, Ankara, Turkey
e-mail: tcakir@ybu.edu.tr

A. Froehlich (ed.), *Spaceports in Europe*, Studies in Space Policy 34,
https://doi.org/10.1007/978-3-030-88311-9_2

In the first scenario, the private initiative developing a micro launcher conducts its operations under a foreign jurisdiction. For instance, German startup Rocket Factory Augsburg, a participation of OHB SE, aims to conduct its maiden launch in 2022 from Andoya Space Center (Norway).[3] Similarly, an agreement has been signed between Isar Aerospace and Andoya Space Center for exclusive access to one of two launch pads (under construction) for 20 years.[4] Isar Aerospace announced another agreement with French Space Agency (Centre National des Etudes Spatiales, CNES) to launch its Spectrum rockets from Guiana Space Center (Centre Spatial Guyanais, CSG).[5] Under this scenario, projects are generally backed by private investors and supported by national government policies outside the European institutional frameworks. It does not mean that European institutions indirectly support the development of privately micro launcher initiatives in Europe. For example, the European Space Agency (ESA), through Boost! initiative, awarded the first contracts to launch startups Isar Aerospace, Hylmpulse and Rocket Factory Augsburg.[6]

In the second scenario, the private initiative developing a micro launcher performs launches under the jurisdiction of its State of origin. British companies aiming to conduct their launches from spaceports located in Scotland is a good example reflecting this scenario.[7] Under this scenario, the launching of a space object is procured by a foreign State. In the first scenario, the procuring State may also be a foreign State.

Even if national backing of micro launcher projects provides certain advantages, securing independent access to space may require an institutional approach in the future.[8] In the third scenario, a micro launcher programme is developed within the framework of European organizations (the ESA or the European Union (EU)). In this respect, the French President Macron has recently proposed a single European micro launcher programme.[9] For the time being, there is not enough determination from the European States on the realization of this scenario. Private competition in the field of micro launchers is generally desired, instead of searching for an institutional approach.

---

[3] Andrew Parsonson, "Germany's launch startups race to grow their launch manifests", (Space News, 27 April 2021) <https://spacenews.com/germanys-launch-startups-race-to-grow-their-launch-manifests/> All websites cited in this publication were last accessed and verified on 15 August 2021.

[4] Andrew Parsonson, "German startup Isar Aerospace signs first launch contract", (Space News, 26 April 2021) <https://spacenews.com/german-startup-isar-aerospace-signs-first-launch-contract/>

[5] "Isar Aerospace will launch small sats from French Guiana", (Spacewatch.global, 15 October 2020) <https://spacewatch.global/2020/10/isar-aerospace-will-launch-small-sats-from-french-guiana/>.

[6] ESPI, Yearbook 2020 (no 1) 102–103.

[7] Ibid 67.

[8] Matteo Tugnoli, Martin Sarret, Marco Aliberti, "European Access to Space: Business and Policy Perspectives on Micro Launchers Executive Summary" (May 2018) ESPI Public Reports 6–7.

[9] Peter B. De Selding, "French president proposes single Euro micro-launcher program, backs EU Commission broadband constellation", (Space Intel Report, 22 February 2021) <www.spaceintelreport.com/french-president-proposes-single-euro-micro-launcher-program-backs-eu-commission-broadband-constellation/>

The CSG will continue to be Europe's main spaceport especially for launches in geostationary orbit. The next generation launch capabilities Ariane 6 and Vega-C will be operated from this spaceport "in the foreseeable future".[10] In the fourth scenario, the established launch provider Arianespace deploys payloads from one of the new spaceports established in Europe.[11] This may be possible with the development of a smaller configuration of Vega even if its political feasibility is low.[12] Indeed, the multiplication of new spaceports would be beneficial to guarantee Europe an autonomous access to space against the possible sudden unavailability of the CSG.[13]

In any scenario, the launch would be qualified as a joint launch involving several launching States. According to Article V(1) of the Convention on International Liability for Damage Caused by Space Objects,[14] the launching States of a joint launch shall be jointly and severally liable for any damage caused by a space object. The term launching State is defined in the first Article of the Convention, and according to this definition it means: (i) a State which launches or procures the launching of a space object; (ii) a State from whose territory or facility a space object is launched. These States are liable for compensation on equal footing and there is no hierarchical order among these States.

The victim of damage caused by a space object can seek the entire compensation from any or all of the launching States. If a launching State has paid the compensation to the victim, this State shall have the right of recourse to other launching States (Article V(2)). Certainly, the multitude of liable States is an example of victim orientation characterization of the space liability régime.[15] At the beginning of the Space Age, the two superpowers intended to assure developing countries concerned about the ultra-hazardous nature of space activities through a protective liability régime in exchange for having the freedom of use of outer space.[16] In this respect, the Liability Convention has found a positive, effective balance between the interests of industrialized countries and the interests of developing ones.

It is essential to remember that the Liability Convention is *lex specialis* of Article VII of the Treaty on Principles Governing the Activities of States in the Exploration

---

[10] Marco Aliberti, Matteo Tugnoli, "The European Launchers between Commerce and Geopolitics" (March 2016) 56 ESPI Report 7.

[11] This scenario has been raised by Arianespace's CEO Stephane Israel: "We could perfectly use other spaceports, but today our spaceport is the Guyana space centre". See Peter Teffer, "Guiana spaceport undaunted by European newcomers", 26 August 2018 <https://euobserver.com/science/142663>.

[12] Aliberti, Tugnoli (no 10) 80, 86.

[13] Ibid 85.

[14] Entered into force 1 September 1972, 961 UNTS 187 (referred to hereinafter as the Liability Convention or the Convention).

[15] Manfred Lachs, *The Law of Outer Space, An Experience in Contemporary Law-Making* (Martinus Nijhoff Publishers, 2010) 114.

[16] See further Marcia Alvarenga dos Santos, "International responsibility and liability of States in the conduct of space activities", Stephan Hobe and Philippe Achilleas (ed.) *50 years of Space Law-Space Law in 50 years* (Brill/Nijhoff, 2020) 231–233.

and Use of Outer Space, including the Moon and Other Celestial Bodies,[17] which establishes the general principle of liability for damage caused by space objects.[18] Article VII of the Outer Space Treaty allocates only third-party liability to the launching State without clarifying the basis of liability, the compensation procedure etc. The Liability Convention, expanding on Article VII of the said Treaty, has established the principle of absolute liability in respect of damage caused by a space object on the surface of the Earth or to aircraft in flight. The victims need not to prove that the fault is that of the launching State. The focus on the victim is also reflected in the limited number of cases of exoneration from the absolute liability provided in the Article VI of the Convention.[19] Another example of victim orientation of the Convention is that State liability is not limited in time, amount, and location.[20]

Sources of Space Law are becoming increasingly diversified.[21] One of the consequences of this diversification has been the proliferation of bilateral agreements which are extending the material scope.[22] The apportionment of financial obligation is generally tackled in bilateral agreements concluded between launching States. According to Article V(2) of the Liability Convention, States may adopt internal agreements ("liability agreements"[23]) to share their high level of liability in relation to joint launches. The conclusion of such agreements would be of utmost importance in the European context given the multiplication of spaceports and micro launcher projects. This contribution addresses how bilateral agreements containing provisions regulating the apportionment of liability have played a fundamental role in space risk management in the European framework (2) before underlining the significance of these agreements within the functioning of the space liability régime (3). Finally, it analyzes the possible impact of bilateral agreements on the elaboration of a coordinated approach in the new European context (4).

---

[17] Entered into force 10 October 1967, 610 UNTS 205 (referred to hereinafter as the Outer Space Treaty).

[18] Armel Kerrest, Lesley Jane Smith, "Article VII", in Stephan Hobe, Bernhard Schmidt-Tedd and Kai-Uwe Schrogl (ed) *Cologne Commentary on Space Law: Volume I, Outer Space Treaty* (Carl Heymanns Verlag, 2009) 129.

[19] See further Bruce Hurwitz, *State liability for Outer Space Activities in accordance with the 1972 Convention on international liability for damage caused by space objects* (Martinus Nijhoff Publishers, 1992) 40–43.

[20] Kerrest, Smith (no 18) 130 footnote 19.

[21] See further Ingo Baumann, "Diversification of Space Law", Marietta Benko and Kai-Uwe Schrogl (ed) *Space Law, Current Problems and perspectives for future regulation* (Eleven International Publishing, 2005) 47–74; Tugrul Cakir, "Les Sources du Droit de l'Espace", Stephan Hobe and Philippe Achilleas (ed.) *50 years of Space Law-Space Law in 50 years* (Brill, 2020) 83–126.

[22] For an analysis in a greater detail on bilateral agreements, see Marco G. Marcoff, *Sources du droit international de l'espace* (1980) 168 Recueil des Cours 54–55; Huang Jiefang, "The Roles of Bilateral Agreements in Space Law" (1988) 31 Proc. on L. Outer Space 172–176.

[23] Committee on the Peaceful Uses of Outer Space, "Review of the concept of the "launching State"", Report of the Secretariat, A/AC.105/768 (January 2002) 12.

2 Liability Agreements in the European Context

2.1 In General

Here are the main examples of agreements apportioning financial obligation in the European context.

The ESA Council's Resolution defines the consequences of the Agency's liability for damages caused during the execution of its space programmes or activities.[24] It is necessary to recall that the ESA deposited a declaration of acceptance of rights and obligations included in the Liability Convention.[25] Therefore, the Agency itself can be considered as a launching State for launches carried out from its facilities at the CSG. As a general principle, the Agency shall indemnify Member States and States participating in ESA space programmes and activities against liability incurred by them in connection with the execution of such programs and activities (A-I-1). If a claim for compensation is addressed to the ESA, the Agency shall conduct the proceedings (A-II-1). If a claim for compensation is addressed to a Member State or a participating State, that State shall consult the ESA without delay (A-II-2). The Agency, if authorized by the applicable law, may join in the proceedings, and substitute itself for the State in question if the latter so requests. The expenditure made by the ESA will be charged to the States participating in the programme concerned, proportionately to their financial contributions to the said programme (A-III-3). Where the risk is covered by insurance, the corresponding premium is charged to the programme. When providing a launch service, the Agency shall take the necessary steps to ensure that the launch service customer takes out an insurance policy covering its liability and that of the Agency for any damage that may result from that service (B-II-1). However, the Agency shall remain liable for damages caused by gross negligence or a deliberate act or omission on its own part or on the part of persons in-its service, unless unanimously decided otherwise by the ESA Council (B-II-2).

Another example of application of Article V of the Liability Convention can be found in the Agreement on San Marco Satellites Tracking and Launching Station

---

[24] Resolution of the Council of the European Space Agency on the Agency's Legal Liability (ESA/C/XXII/RES.3, 13 December 1977). This Resolution may be considered as an important contribution in the development of Space Law. For further information on the relationship of the ESA with Space Law see Frans von der Dunk, "European Space Law", Frans von der Dunk, Fabio Tronchetti (ed.) *Handbook of Space Law* (Edward Elgar Publishing, 2015) 223–228.

[25] Committee on the Peaceful Uses of Outer Space Legal Subcommittee Sixtieth session Vienna, 31 May–11 June 2021, Status of International Agreements relating to activities in outer space as at 1 January 2021, A/AC.105/C.2/2021/CRP.10 (May 2021) 10.

concluded between the ESA, Italy, and Kenya in 1995.[26] In this agreement,[27] Kenya authorized the ESA to use the Malindi Station for supporting Ariane launches (Article 1.1). Article 10 of the Agreement provides for a liability sharing mechanism. As a general principle, Kenya would not be held liable for the activities of the ESA in its territory (Article 10.1). If the liability is that of Kenya, it shall have a right of recourse against the ESA (10.2.). Furthermore, the ESA shall hold Kenya harmless if any suit, action, or claim is brought against it in respect of ESA activities at the Malindi Station (10.3.). The second and third paragraphs are not applicable if the damage arises out of gross negligence, an act or deliberate omission of the Kenyan Government or a person acting on its behalf.

In British practice, we can also find examples of liability agreements. In an exchange of notes between the United Kingdom and the People's Republic of China on the apportionment of liability arising during the launch phase of Asiasat, the latter shall assume the liability and indemnify the former if its liability arises (Article 3).[28] In the event of a claim against the United Kingdom, it shall notify the People's Republic of China as soon as possible (Article 4(1)). The United Kingdom shall not complete settlement with any claimant without full consultation with the People's Republic of China (Article 4(2)). If the latter does not agree with the terms of a proposed settlement between the former and a claimant State, the UK shall submit the claim to a claims commission (Article 4(3)). Concerning the liability arising from the launch phase, other agreements have been passed between the two States, using

[26] Protocol between the European Space Agency, the Government of the Republic of Italy and the Government of the Republic of Kenya on the setting up and operation of European Space Agency equipment within the perimeter of the San Marco Satellites Tracking and Launching Station in Malindi, Kenya, and on the cooperation between the Government of the Republic of Kenya and ESA for peaceful purposes, 13 September 1995. The San Marco Project was crucial at the beginning of the Italian space research programme in the 1960s. This project was developed in close cooperation with NASA. On the origins of this programme see generally H. Nesbitt, "History of the Italian San Marco Equatorial Mobile Range" (1971) NASA CR-111987 <http://ntrs.nasa.gov/archive/nasa/casi.ntrs.nasa.gov/19720007316.pdf>. The project had two main segments. The sea segment comprised two fixed installations on the High Seas: San Marco and San Rita. These installations are no longer operational. The land segment is Malindi Tracking Station, which is still operational and serves for tracking ESA, NASA, and Italian satellites. On the implementation of this project and the Italy-Kenya space cooperation, see further Ornella Ferrajolo, "Launch and tracking stations: The "SanMarco-Malindi" Case", Gabriel Lafferanderie (ed.) *Outlook on space law over the next 30 years* (Kluwer Law International, 1997) 273–284.

[27] On cooperation agreements concluded by ESA and the origins/application of Article XIV of the ESA Convention, see further Catherine Baudin, "Cooperation and International Agreements, Article XIV of the ESA Convention" (1998) XXIII Air & Space Law 8–15.

[28] Exchange of Notes between the Government of the United Kingdom of Great Britain and Northern Ireland and the Government of the People's Republic of China concerning Liability for Damage arising during the Launch Phase of the Asiasat Satellite, Peking, 26 March 1990 and 2 April 1990. This exchange of notes constitute an agreement between the two States.

almost identical terms.[29] An agreement has also been concluded between the United Kingdom and the Republic of Chile in the same manner.[30]

A liability scheme has also been developed within the framework of European launcher programme. It should be noted that France is a launching State for any launch carried out from the CSG or conducted by Arianespace outside the CSG. Arianespace's operations may give rise to the liability of France, the ESA, its Member States and/or the Russian Federation.

2.2 *Liability Agreements in the Framework of European Launcher Programme*

It is useful to examine the internal liability sharing agreements in the light of the launch vehicles: Ariane (heavy), European version of the Soyuz (medium), and Vega (light).

2.2.1 Ariane

General Context

According to the French president De Gaulle, space activities were an important way for the French to prove their sovereignty and strategic independence in a new technological field.[31] In this respect, security and defense considerations were the driving forces behind the development of space technologies in France. Instead of developing national capacities, France has pragmatically decided to realize its objectives in the European multilateral context.[32] Certainly, one of the successes of the French Space

[29] Exchange of Notes between the Government of the United Kingdom of Great Britain and Northern Ireland and the Government of the People's Republic of China concerning Liability for Damage during the Launch Stage of the Apstar-1, Apstar-2, and Asiasat-2 Satellites, Peking, 28 June 1994; Exchange of Notes between the Government of the United Kingdom of Great Britain and Northern Ireland and the Government of the People's Republic of China concerning the Liability for Damage during the Launch Stage of the Apstar-IA and Apstar-IIR Satellites, Peking, 5 February 1996 and 6 February 1996.

[30] Exchange of Notes between the government of the United Kingdom of Great Britain and Northern Ireland and the government of the Republic of Chile concerning liability for damage during the launch phase of the Fasat-Bravo Satellite, 30 June 1999.

[31] See further on the origins of French Space Programme and France's contribution to the development of European launch capabilities Hervé Moulin, *La France dans l'Espace 1959–1979* (Division des Publications de l'Agence spatiale européenne ESA/ESTEC, June 2006).

[32] Conseil d'État, *Pour une politique juridique des activités spatiales* (2006) 21.

Strategy has been its capacity to find a balance between its goal of independence and the promotion of international cooperation since the presidency of De Gaulle.[33]

On the European side, similar awareness of having an independent launch capability has emerged since 1960s.[34] The strategic and technologic model founded by Europe in 1980s has been mainly based on the principle of non-disassociation between sovereignty and competitivity, which ensured it a privileged position in the market until the 2010s.[35] Challenges posed by the insufficient number of institutional launches in Europe and the growing international competition constitute real dangers to this foundational principle. In the light of this, Europe whose leadership is "eroding"[36] in launchers should reconsider the initial balance found in the 1980s and develop a more realistic and coherent policy.

Ad Hoc International Agreements

Examples of agreements concluded in the framework of Ariane programme include:

A framework agreement concluded on 5 May 1976 specified the relationship between ESA /the French Government and their mutual rights and obligations with respect to the CSG. This agreement has been regularly renewed. Article 17 of the Agreement, signed on 18 December 2008,[37] provided a risk-sharing mechanism between the French government and the ESA:

(1) Launches carried out in the framework of the ESA's development programmes: The Agency assumes liability in accordance with the ESA Council's resolution (mentioned above) and shall hold the French Government harmless from any claims made against it for any damage, as defined in the said resolution (Article 17-1.1.).

---

[33] Jean-Luc Lefebvre, Guilhem Penent, "La politique spatiale de la France: Une vision gaullienne pour l'Europe" (2013) 62 ZLW 98. It is worth noting that the German expertise in propulsion systems was also crucial in the success of Ariane launchers. Alaine Dupas, Stéphane Janichewski, Wulf von Kries, Kai-Uwe Schrogl, "A Franco–German view of Europe's ambition in space for the twenty-first century" (2001) 17 Space Policy 104. "Europe's autonomy and presence in space is essentially the result of the two countries' complementary engagement." Ibid.

[34] See further Anna Clementina Veclani, Jean-Pierre Darnis, "European Space Launch Capabilities and Prospects", Kai-Uwe Schrogl (ed.) *Handbook of Space Security. Policies, Applications and Programs Volume 2* (Springer, 2014) 786.

[35] See further Bruno Erin, "L'émergence des acteurs privés dans l'espace extra-atmosphérique et les conséquences pour l'Europe" (2016) XVII Annuaire Français de Relations Internationales 819–831. Similarly, Guilhem Penent highlights four foundational tensions for the European launcher initiative, which are also determinant for future: sovereignty and rentability, commercial and institutional, private and public, European and national. Guilhem Penent, "Ariane 6: les défis de l'accès à l'espace en Europe" (November 2014) Actuelles de l'IFRI 5.

[36] ESA, "Agenda 2025, Make Space for Europe" (2021) 14.

[37] L'accord entre le Gouvernement de la République française et l'Agence spatiale européenne relatif au centre spatial guyanais et aux prestations associées (ensemble deux annexes), signed in Paris on 18 december 2008, (referred to hereinafter as the CSG Agreement).

(2) Ariane launches operated by Arianespace from the CSG: The French Government shall, vis-à-vis the Agency and its Member States, pay any damages that may be awarded (Article 17-1.2.).
(3) Launches carried out in the framework of French national space programme: The French Government shall hold ESA and its member States harmless from any claims (Article 17-1.6.).
(4) Other launches: The apportionment of liability is governed under ad hoc agreements concluded between the ESA, France, and the concerned State (Articles 17-1.7)
(5) Exceptions: The hold harmless agreements are not applicable if the damage arises out of gross negligence, an act or deliberate omission of the Government of the French Government or of the Agency, of persons acting on its behalf and of its member States (except for the French State) (Article 17-1.5.).

Another key instrument in the liability apportionment is the Declaration by Certain European Governments Relating to the Ariane Launcher Production Phase adopted on 14 January 1980.[38] Through this Declaration, the execution of the production phase of the Ariane launchers has been entrusted to the Arianespace, a company created under French Law (1.3.). The Declaration has put in place a liability régime applicable in case of damage caused by launches operated by the company.[39] Under this régime, in the event of recourse by the victims of damage caused by Ariane launches, the French Government shall be responsible for the payment of any damages awarded (4.1.). In this case, Arianespace shall reimburse the French Government within a ceiling of 400 million French francs (60 million euros) per launch (3.8.). The operator shall benefit from the guarantee of State from that limit upwards.[40]

The Arianespace declaration has been regularly renewed. The Launchers Exploitation Declaration adopted on 30 March 2007, amended on 4 December 2017, has renewed, and prolonged it until the end of 2035.[41] The object of this Declaration is to establish a common framework for the exploitation phase of Ariane, Vega, and Soyuz launchers (Article I.1.). Article IV of the Declaration reaffirms and clarifies provisions of the CSG Agreement and the Arianespace Declaration with respect to liability apportionment. The principles set forth above have also been transposed into

[38] Declaration by Certain European Governments Relating to the Ariane Launcher Production Phase (hereinafter referred to as the Arianespace Declaration), 14 April 1980. For further information see Michel G. Bourély, "La Production du Lanceur Ariane" (1981) 6 Annals of Air and Space Law 279–314.

[39] For a more detailed assessment, see Julien Hermida, "Risk Management in Arianespace Launch Agreements" (2000) 25 Annals Air & Space L. 143–155.

[40] See further Tugrul Cakir, "La garantie de l'Etat dans la loi française relative aux opérations spatiales: un atout de la politique juridique de la France ?" (2017) 66 ZLW 104–112.

[41] The Declaration by certain European governments on the launchers exploitation phase of Ariane, Vega, and Soyuz from the Guiana Space Centre (CSG) adopted on 30 Mars 2007 and amended on 4 December 2017.

the French Space Operations Act (FSOA).[42] In doing so, ad hoc practices have been generalized and greater legal certainty has been provided for private entities.[43]

The access to the CSG is governed by Article 10 of the CSG Agreement. This requires the agreement of the French Government and ESA (para.1) and the conclusion of agreements to define, in particular, the provisions applicable to liability (para.3). It is in this regard that the CSG has been opened to Soyuz.

2.2.2 Soyuz-ST

General Context

Soyuz, considered as a reliable launcher since the 1960s, has been designed for medium size missions. In this respect, Soyuz at the CSG programme has allowed Europe to launch all the range of spacecraft sizes and to augment the competitiveness of the European launcher industry on the global market. On the Russian side, the main benefit has been the improved access to markets, the enhanced performance of the launcher, thanks to the geographical advantages offered by Kourou, and the reduction of its dependency on the Baikonur Space Center.[44]

The space cooperation between Russia and Europe has been maintained even under difficult political circumstances.[45] It is especially worth noting that Europe has relied on Russian launchers for two decades.[46] One of the main goals with Ariane 62 (medium-lift launcher) is to replace Soyuz-ST, considered as a "transition solution", and reduce dependency on this rocket.[47] With the arrival of Ariane 62, the future of the Euro-Russian partnership in the field of launchers seems to be uncertain.[48] In any event, Soyuz at the CSG programme is one of the most important examples of space cooperation in the field of launchers.

---

[42] Loi n° 2008–518 du 3 juin 2008 relatives aux opérations spatiales published in the French Official Gazette (JORF) on 4 June 2008. See further on a detailed analysis of the FSOA Mireille Couston, "La loi sur les operations spatiales" (2009) 58 ZLW 253–282. Most of the latest agreements concluded in the framework of the European launcher programme have been submitted to the French Parliament for ratification to "give them an enforceable legislative value". Philippe Clerc, *Space law in the European context: national architecture, legislation and policy in France* (Eleven International Publishing, 2018) 118 footnote 66, 122.

[43] Bernhard Schmidt-Tedd and Isabelle Arnold, "The French Act relating to space activities-From international law idealism to national industrial pragmatism", 11 ESPI Perspectives (2008) 3.

[44] Aliberti, Tugnoli (no 10) 17; Clerc (no 42) 118–119.

[45] Florian Vidal, "Russia's Space Policy: The of Decline?" (January 2021) Etudes de l'IFRI 40–41.

[46] Ibid.

[47] Marco Aliberti, Arne Lahcen, "The Future of European Flagship Programmes in Space" (2015) 53 ESPI Report 50–51.

[48] See further on the prospects for European and Russian cooperation in space Eva Avrillon, "European-Russian Space Cooperation in 2030: From Commercial Partnerships to Common Exploration Programmes?" (January 2013) 65 ESPI Perspectives.

Ad Hoc International Agreements

The implementation of the Soyuz at the CSG programme has led to the conclusion of bilateral agreements between France, Russian Federation, and the ESA.

France-Russian Federation cooperation in the field of space has mainly concentrated on research and technology. A cooperation framework agreement was signed on 26 November 1996 between these States.[49] The aim of this agreement is to develop a closer partnership in all aspects of space (scientific, technical, industrial, and commercial) (Article 1). Launch systems and services are one of the four fields of bilateral cooperation (Article 2). As a result of this, Starsem has been created in 1996 as a Russian/French joint venture to commercialize Soyuz launch vehicles from Baikonur Space Center. The success of the Starsem Project has given rise to the birth of the Soyuz at CSG programme.

An agreement between France and the Russian Federation authorizing the implementation of Soyuz from the CSG was signed on 7 November 2003.[50] The purpose of the Agreement is to define general framework of the use of Soyuz at the CSG (i.e. national space agency's responsibilities and liabilities).[51] Article 10(2) of the Agreement provides that the compensation of victims of damage caused by a Soyuz launch shall be made on a basis of parity (*partage paritaire*).[52] The same Article also limits the liability of Arianespace to the amount of 60 million euros for each launch performed from the CSG.[53]

Another fundamental agreement in the implementation of Soyuz from the CSG was signed on 19 January 2005 between ESA and Roscosmos.[54] One of the main objectives of the agreement was to define and clarify the responsibilities of the parties concerning the exploitation of Soyuz (the other being research and development for future launchers).[55]

The agreement between the ESA and France signed on 21 March 2005 clarified the relationship between the French Government and the Agency concerning the Soyuz Launch Facility (SLF) at the CSG and the conditions of guarantee that the French Government granted to the ESA and its Member States (Article 2).[56] In the event of a

---

[49] L'Accord entre le Gouvernement de la République française et le Gouvernement de la Fédération de Russie relatif à la coopération dans le domaine de l'exploration et de l'utilisation de l'espace à des fins pacifiques (ensemble une annexe), signed in Paris on 26 November 1996.

[50] L'Accord de coopération à long terme dans le domaine du développement, de la réalisation et de l'utilisation de lanceur l'implantation du lanceur Soyouz-ST au CSG, signed on 7 November 2003.

[51] Clerc (no 42) 119.

[52] Conseil d'Etat (no 32) 48.

[53] Ibid 56.

[54] L'Accord entre l'ASE et l'Agence spatiale fédérale russe (ROSCOSMOS) relatif à la coopération et au partenariat à long terme dans le domaine du développement, de la réalisation et de l'utilisation des lanceurs, signed on 19 January 2005.

[55] European Commission CORDIS, "ESA and Russia launch new partnership" (2005) < https://cordis.europa.eu/article/id/23212-esa-and-russia-launch-new-partnership >

[56] L'accord entre le Gouvernement de la République française et l'Agence spatiale européenne relatif à l'Ensemble de lancement Soyouz (ELS) au Centre spatial guyanais (CSG) et lié à la mise

claim addressed to one of the Parties, Parties to the agreement shall consult each other as soon as possible (Article 12(1)). In the framework of the commercial exploitation of the Soyuz programme at the CSG, the French Government shall hold the ESA and its member States harmless from any claims relating to damages caused due to the execution of launch activities at the CSG (Article 12(3)). However, the hold harmless agreement would not be applicable if the damage arises out of gross negligence, an act or deliberate omission on the part of the Agency, of persons acting on its behalf or of its Member States (except for the French State). These principles have been reiterated in the CSG Agreement (Article 17.1.4) which has replaced the agreement concerning the SLF, and in the Launchers Exploitation Declaration (Article IV(c)).

2.2.3 Vega

General Context

The birth of the Vega project, as in the case of Soyuz, arose from the ESA's desire to diversify European launch capabilities in the late 1990s.[57] The project has been conceived as a European answer to the emergent need to deploy micro-satellites and to complement the other European launchers. Italy, an important player in the launcher field along with France and Germany, has taken a leadership role in its development since 1998. Vega-C has been developed to improve the performance of the Vega launch system and to be more competitive in the market.[58] The maiden flight of Vega-C is scheduled for the second half of 2021.

Ad Hoc International Agreements

Under Article 17 of the CSG Agreement, in the event of recourse due to damage caused by a Vega launcher from the CSG during the exploitation phase, the French Government shall, vis-à-vis the Agency and its Member States, be responsible for the payment of one third of damages and the Agency being obliged, vis-à-vis the French Government, to pay the remaining two thirds. The liability of the French Government has been repeated in the Launchers Exploitation Declaration (Article IV(b)). Concerning the ESA's liability, Member States having participated in the relevant development programmes shall conclude an agreement regulating the division of any financial burden. Conversely, Arianespace shall reimburse the French government and the ESA, pro rata to their respective shares of liability and within a ceiling of 60 million euros per launch (Article 3.1.j.).

---

en œuvre du programme facultatif de l'Agence spatiale européenne intitulé 'Soyouz au CSG' et à l'exploitation de Soyouz à partir du CSG (ensemble deux annexes), signed in Paris on 21 March 2005.

[57] Aliberti, Lahcen (no 47) 47.

[58] Aliberti, Tugnoli (no 10) 29.

After examining how Article V(2) of the Liability Convention has been implemented in the European context, we will focus on the importance of liability agreements within the functioning of the space liability régime.

3 Liability Agreements in Space Liability Régime

3.1 Liability Agreements Complement the Liability Convention as an Important Source of Space Law

It is worth noting that the liability régime put in place by the Liability Convention is not complete, but reinforced by other sources of law: national space legislations[59] and ad hoc agreements. These sources facilitate the implementation of the Convention and overcome its limitations. To give an example, the Convention does not include provisions regarding the share of financial obligation except in Article IV, which provides for the internal sharing of the burden of compensation between launching States involved in an on-orbit accident.[60] Therefore, in other cases of joint and several liability, these agreements should be conceived as an adapted means of dealing with the division of financial burden in the light of the characteristics of space operation.

These agreements serve to avoid potential problems concerning risk distribution.[61] They provide legal certainty and clarification against the risks implicated in the operation. If no liability agreement is concluded, the reaching of such an agreement could be difficult after the liability of a launching State has arisen.[62] Even worse, if no agreement is concluded, the launching State paying the whole compensation, risks not being able to obtain the recovery:

> The party found liable in law may be left to assume the claim in full, without any subsequent internal breakdown of the financial burden with the joint launching States. The inability to further subdivide the obligation to compensate may occur for various reasons, such as the

---

[59] International obligations stemming from space treaties are not binding for private entities, but States shall ensure that the activities of these entities comply with their international obligations. Certainly, Article VI of the Outer Space Treaty does not impose an obligation to enact a domestic legislation. However, doing so, permits States to conserve their national interests and encourage commercial space activities. Through different means, States find solutions to lighten their burdens posed by the liability régime.

[60] See further Lesley Jane Smith, Armel Kerrest, "Liability Convention", Stephan Hobe, Bernhard Schmidt-Tedd and Kai-Uwe Schrogl (ed) *Cologne Commentary on Space Law: Volume II* (Carl Heymanns Verlag, 2013) 140.

[61] Hurwitz (no 19) 38.

[62] Matxalen S. Aranzamendi, Frank Riemann, Kai-Uwe Schrogl, "The 2004 Resolution on the Application of the concept of the 'Launching State'", Stephan Hobe, Bernhard Schmidt-Tedd and Kai-Uwe Schrogl (ed) *Cologne Commentary on Space Law: Volume III* (Carl Heymanns Verlag, 2015) 388.

> lack of financial liquidity, where one of the State parties is unable to fulfill its portion of the financial burden. It may also arise where the other wrongdoer cannot be traced.[63]

Ad hoc agreements are useful to share liability among launching States in advance, and they are not a danger to the coherence of the space liability régime. Through the conclusion of these agreements, States cannot amend the provisions of the Liability Convention.

3.2 The International Space Liability Régime Cannot Be Altered by Liability Agreements

Liability agreements deal with inter-parties-liability. Therefore, the interests of the victim are safeguarded as they are not bound by the content of this type of accords and may seek the entire compensation from any or all the launching States.[64] In this respect, it would be suitable to have a specific example.

Article 8.4. of Russia-Kazakhstan Treaty on the Leasing of the Baikonur Complex stipulates that "in the event of damage connected to the Baikonur cosmodrome activity in performance of the Russian space programs, only Russia bears liability as the Launching State according to the 1972 Convention on International Liability".[65] In the light of this provision, the Republic of Kazakhstan shall not be regarded as a launching State if damage is caused in the context of Russian space programmes. The only exception to this is the joint launch conducted by these two States; their liability is joint and several in this case.[66] However, as underlined above, a third party is not bound by the provisions of this Treaty and may seek compensation from Kazakhstan, which shall be regarded as a participant in all joint launches conducted from Baikonur Space Center.

The space liability régime may be found to be strict by States which have a limited (or no) participation in the conduct of space operation. These States have an interest in concluding liability agreements to protect themselves against the potential for a high level of liability.

---

[63] Smith, Kerrest (no 60) 144.

[64] The choice of the claimant State is unsurprisingly connected to "its ability to satisfy the compensation claim in full". Ibid 145.

[65] Gennady P. Zhukov, "Can the State from Whose Territory a Space Object Was Launched Declare Itself a Non-Launching One?" (2003) XXVIII Air & Space Law 51.

[66] Ibid.

3.3 *Liability Agreements Are Essential Tools Against the "Unfairness" Caused by the Liability Convention*

In the context of privatization and commercialization of space activities, the launching State concept has been challenged.[67] As underlined by many scholars, holding liable States which have nothing to do with the control of the space object, may be seen "unfair".[68] This could also have a dissuasive effect on States intending to host a foreign launch facility on their territory and undermine the development of international cooperation.

In this context, the launching State concept has generally been challenged by States from whose territory a space object is launched. According to Article V(3) of the Convention, a State from whose territory or facility a space object is launched is considered to be participant in a joint launch. Indeed, this paragraph is a reaffirmation of the first Article of the Convention defining a 'launching State'.[69] This repetition may be explained by the controversial nature of the liability of the State of the territory of launch.[70] Therefore, these States cannot deny their liability and they are liable to the same degree as the other launching States. The rigidity of the space liability régime should be considered as an incentive for States to take necessary precautions before allowing a State to install launch facilities on their territory. The conclusion of an agreement referred to in Article V(2) of the Liability Convention would be one of the appropriate precautions to take.

The launching State concept has also been challenged by States providing launch services. In the view of certain States, the liability for damage caused by a payload during the orbital control phase should be that of the State which operates it.[71] Those States would be liable only for damages caused by the launch vehicle, as they

---

[67] Irmgard Marboe, "The 2013 Resolution on recommendation on national legislation relevant to the peaceful exploration and use of outer space", Stephan Hobe, Bernhard Schmidt-Tedd and Kai-Uwe Schrogl (ed) *Cologne Commentary on Space Law: Volume III* (Carl Heymanns Verlag, 2015) 499.

[68] Hurwitz (no 19) 38; Lotta Viikari, "A New Liability Regime for the Space Sector - An Economic Imperative" (2010) 3 Indian J. Int'l Econ. L. 119; Armel Kerrest, "D'un droit interétatique issu de la guerre froide à l'encadrement des activités spatiales" Armel Kerrest (ed.) *Le droit de l'espace et la privatisation des activités spatiales* (Pédone, 2003) 15.

[69] Smith, Kerrest (no 60) 146.

[70] See for example: "Liability should not attach to States providing territory for the launching of space objects. For should liability so attach, States would be discouraged from providing territory for this purpose and land-locked countries may encounter difficulties in obtaining facilities from which their space objects might be launched." UN Doc. A/AC.105/21, Report of the Legal Sub-Committee on the Work of the Second Part of its Third Session (5–23 October 1964), 23 October 1964, Annex IV, Summary of views expressed in Working Groups I and II, Summary of points raised in discussions of Working Group II, Part I, Points raised in general discussion, p. 8.

[71] See for instance: "Some countries have expressed the view that the State or States providing launch services should not be liable for damage caused by a payload after the payload has been placed successfully into the proper orbit. Thereafter, in the view of those countries, the State or States that own or operate the payload should be liable for damage caused by it." Committee on the Peaceful Uses of Outer Space, "Review of the concept of the "launching State"" (no X) para. 42.

have no means to avoid the damage caused by the payload placed successfully in orbit. Indeed, through the conclusion of liability agreements, States may introduce differentiation between the launching phase and the operational phase, as shown above. Given that the amendment of the Liability Convention would be difficult to accomplish, introducing such a differentiation would be more pragmatic.

Liability agreements should be regarded as solutions to the challenges posed by the implementation of the launching State concept in the context of commercialization of space activities. The multitude of launching States is a good example of victim orientation of the Liability Convention and this orientation must be maintained. Launching States may conclude liability agreements as desired without endangering this orientation. As space technology matures, States will perceive the need to amend the Liability Convention. The conclusion of liability agreements does not in any way mean the amendment of its provisions. Given the strict nature of liability rules, these agreements especially protect States having limited (or no) participation in the conduct of an operation, and provide them with clarity and certainty. They serve to elaborate and adapt the liability régime to the necessities of the operation, in the light of political and technical realities.

3.4 Liability Agreements in the New European Context

Currently, most space objects have more than one launching State. In this respect, liability channeling mechanisms are becoming more important with respect to the adequate application of space treaties.[72] Following this logic, the Launching State Resolution recommends the conclusion of liability agreements in relation to joint launches and cooperation programmes.[73]

In practice, these agreements are frequently concluded as part of long-term cooperation strategy (especially in the context of exploitation of launch sites and launchers).[74] In the first scenario mentioned above, the launch service provider State and the State of the territory of launch, have an interest to reach such an agreement. If the fourth scenario occurs, France would also conclude a liability agreement with the State of the territory of launch.

In the agreement concluded between launch provider State and the State of the territory of launch, the latter should be protected if it has nothing to do with launch operations. It means that this State would not be held liable through the activities conducted on its territory and that a launch service provider State would hold it harmless from any claims made against it. As we have seen in many examples

[72] Aranzamendi, Riemann, Schrogl (no 62) 387.

[73] Resolution adopted by the General Assembly, A/RES/59/115 of December 2004, Application of the concept of the “launching State”. See further on negotiation and drafting history of paragraph 2 of the Resolution Aranzamendi, Riemann, Schrogl (no 62) 387–388.

[74] See further Tugrul Cakir, “Le recours aux accords internationaux dans le cadre d’une opération spatiale” (2020) 69 ZLW 493–499.

above, the hold harmless agreement would not apply if the damage caused was due to the gross negligence of the State of the territory of the launch. Nevertheless, if the State of the territory of launch is also involved in the conduct of operations, this State would assume a part of the liability in respect of its level of involvement. In this sense, the approach taken in the Agreement passed between France and Russian Federation in 2003 can be used as an example.

The liability sharing mechanism developed by European States in the context of the European launch programme achieves the right balance between all parties involved: the significant part of the financial burden being assumed by France, as shown above. This mechanism strengthened by the transposition of agreements into the FSOA has also provided ideal conditions for Arianespace in respect of redistribution of risks for launch services. If the third scenario occurs, European States would certainly develop an efficient risk management mechanism thanks to the experience obtained from past implementations.

The number of agreements referred to in Article V(2) of the Liability Convention is limited in the context of international organizations. The EU[75] has not yet declared acceptance of rights and obligations provided for in the space treaties other than the Outer Space Treaty.[76] After the deposit of such an acceptance, the issue of legal liability of the EU and the modalities of distribution of financial burden among Member States would be tackled. In this respect, the ESA Council's Resolution on legal liability may give inspiration to the EU and other international organizations. It is also necessary to recall that the implementation of Article VI of the Outer Space Treaty (authorization and supervision of private space activities) by the EU is rather difficult given the ambiguous nature of the provisions of the Lisbon Treaty on the space competence and the complicated issue of implementation of international obligations of Member States at European level (mainly Member States' responsibilities).[77]

Outside the context of exploitation of launch sites/launchers, these concluded agreements are limited in number. Despite this fact, in the second scenario mentioned above, the conclusion of these agreements may be developed in the European context. As shown above, we can point out that these agreements focus on the liability arising from the launch phase. In the European framework, the orbital control phase may also be encompassed in such an agreement. If a claimant State demands compensation

[75] On the involvement of the European Community/Union in Space Activities see generally Frans von der Dunk, "European Space Law" (no 24) 244–250.

[76] As at 1 January 2021, four international organizations, namely the ESA, the European Telecommunications Satellite Organization, the European Organization for Exploitation of Meteorological Satellites and, Interspoutnik made a declaration accepting rights and duties provided for in the Liability Convention. Committee on the Peaceful Uses of Outer Space Legal Subcommittee (no 25) 10.

[77] Frans von der Dunk, "Article VI of the Outer Space Treaty in the European Context" (2008) 51 Proc. Int'l Inst. Space L. 558–559; Bernhard Schmidt-Tedd, "Authorisation of Space Activities after the Entry into Force of the EU Reform Treaty", Frans von der Dunk (ed) National Space Legislation in Europe (Martinus Nijhoff publishers, 2011) 308–310.

from the launching State that has no control over the operation, the State exercising control shall assume the liability and hold the other State harmless from any claims made against it for any damage.

Regarding the first and second scenarios, it would be appropriate to establish clear and efficient guidelines to facilitate the conclusion of these agreements in the European framework. This would greatly benefit the development of the space sector in Europe. The European practice developed in this sense may inspire other countries to expand their legal policy. Agreements elaborated in the European context may also become the foundation on which multilateral treaties are concluded in the future.[78]

A reinforced application of the provisions of space treaties requires a coordinated approach. In this sense, ad hoc agreements (including, probably, provisions on the apportionment of liability between liable States) may ensure coordination between national space legislations and facilitate their implementation. In so doing, States can overcome the limitations of their national space legislations. This demonstrates the complementary relationship between secondary sources of Space Law in the service of space treaties.

4 Conclusion

The emergence of non-governmental entities in the space field has brought about the multiplication of national space legislations, which have been the economic and strategic tools for States. States have enacted national legislation not only to comply with their obligations under the space treaties, but also to elaborate a legal policy, which serves the development of local space industry. One can define "space legal policy" as the strategy employed by a State to balance its international obligations with the expectations of its private sector.[79] The legal policy of States is "interdependent", and each mode of interdependence (competitive, coordinative, informational) has consequences regarding the governance of space activities.[80] The coordinative mode requires coordination among States for authorization and monitoring mechanisms. In this respect, two examples attract our attention.

Article 4(4) of the FSOA provides a simplified authorization procedure for applicants aiming at performing a space operation outside French jurisdiction.[81] In this case, the operator may be exempted from all or any part of the technical compliance procedure. The condition is that "the national and international commitments made

---

[78] See further on initiation function of bilateral agreements in connection with multilateral treaties Jiefang (no 22) 174.

[79] See further Tugrul Cakir, "Les législations nationales relatives aux opérations spatiales comme concrétisation d'une politique juridique" Clémentine Bories, Marina Eudes, Lucien Rapp and Lukas Rass-Masson (ed.) *Droit de l'espace extra-atmosphérique-Questions de l'actualité*, (Presses de l'Université Toulouse 1 Capitole, 2021) 15–30.

[80] See further David Lazer, "Global and Domestic Governance: Modes of Interdependence in Regulatory Policymaking" (2006) 12 European Law Journal 455–468.

[81] See further Clerc (no 42) 159,193,200.

by [the] State [exercising its jurisdiction over the operation] as well as its legislation and practices include sufficient guarantees regarding the safety of persons and property and the protection of public health and the environment, and liability matters."[82] The applicant has the burden of proof on these guarantees.[83] However, the control of conformity of sufficient guarantees should be made through bilateral agreements (dealing also with the internal sharing of liability) given that the control of activities performed abroad is not feasible by French authorities.[84]

Under Section 13(4) of the British Act, a license, authorization, or approval granted by a foreign State may be accepted or recognized by the regulator in determining conditions of licenses.[85] For this purpose, the Secretary of State must from time to time publish a list of States granting licenses that may be taken into consideration (Section 13(5)). A bilateral agreement concluded between the UK and the State granting authorizations could be useful to accelerate the authorization process in the former on a long-term basis.

Taking into account foreign rules is a pragmatic and flexible approach and contributes to the advancement of national authorization procedures (especially concerning technical safety evaluation). Cross border recognition of authorizations would facilitate the authorization process for applicants in terms of time and cost. These solutions are particularly adapted in the European context and would contribute to the development of micro launcher projects and spaceports in Europe.

Tugrul Cakir is a lecturer specialized in Space Law at the Ankara Yildirim Beyazit University since March 2020. He earned a master's degree (2014) and a Ph.D. degree (2019) at the Université Jean Moulin Lyon III. His master's thesis and PhD thesis treated different aspects of International and National Space Law. His research outputs, among others, include a chapter relating to sources of Space Law published in December 2020 by the Hague Academy of International Law in a collected work (*50 Years of Space Law-Space Law in 50 Years*).

---

[82] A non-official English translation of the FSOA can be found in ibid 403–417.

[83] Ibid 200.

[84] Armel Kerrest and Frans von der Dunk, "Liability and Insurance in the Context of National Authorisation", Frans von der Dunk (ed) National Space Legislation in Europe (Martinus Nijhoff publishers, 2011) 157; Couston (no 42) 264.

[85] Space Industry Act, Chaps. 5 and 15 March 2018. See also recommended practice by Project 2001 Plus: "An authorisation should not be required for activities authorized by another state if such an authorisation is granted under comparable requirements and conditions as set up in this act and deem securing compliance with international obligations existing with regard to space activities." Michael Gerhard and Kristina Moll, "Perspectives for more national space legislation-Introduction by rapporteurs", Stephan Hobe, Bernhard Schmidt-Tedd, Kai-Uwe Schrogl (ed) *'Project 2001 Plus' – Global and European Challenges for air and space law at the edge of the 21st Century* (Carl Heymanns Verlag, 2006) 31.

The "European New Space": Spaceports Provided by Private Actors

Stephanie Stipsits

Abstract The involvement of private actors in spaceport services has proved to be one of the driving forces of the *European New Space* dynamic. The undefined terms "non-governmental entity" and "activity taking place in outer space" in the Outer Space Treaty bedevil the assessment of spaceports in international space law. States have already turned to national laws to create a differentiation between vertical-launch spaceports and horizontal-launch spaceports. This article argues for a broad scope of the terms "non-governmental entity" and "activity taking place in outer space" in Article VI Outer Space Treaty. It concludes that ownership structures as well as the specific functions of the entity are necessary to identify the role of private actors as "non-governmental entity". It also argues that States bear international responsibility for any kind of spaceport activity, irrespective of the status as vertical- or horizontal-launch spaceport. In particular, States have to authorise and continuously supervise the spaceports of non-governmental entities. This chapter also posits that the current international space law regime on liability applies to spaceports.

1 Introduction: The Space Industry on the Advance

At the beginning of human space activities, the utilisation of outer space was dominated by the public domain. States considered outer space relevant for their interests and conducted space activities for security and military reasons.[1] The potential of outer space goes beyond strategic considerations, and commercial activities soon claimed their presence in outer space. Satellite telecommunications and remote sensing are indispensable for our daily lives and represent an advantage of the use of outer space. The economic significance of outer space attracted private actors,

[1] Stephan Hobe, 'Historical Background' in Stephan Hobe, Bernhard Schmidt-Tedd and Kai-Uwe Schrogl (eds), *Cologne Commentary on Space Law, Volume I* (Carl Heymanns Verlag 2009) 10.

S. Stipsits (✉)
Department of European, International and Comparative Law, University of Vienna, Vienna, Austria
e-mail: stephanie.stipsits@univie.ac.at

A. Froehlich (ed.), *Spaceports in Europe*, Studies in Space Policy 34,
https://doi.org/10.1007/978-3-030-88311-9_3

and investments led to a new drive in the space industry. This increased engagement of private actors in the space sector is part of a "New Space" dynamic, where the supremacy of the public domain is increasingly complemented by the private domain. This trend of private entities dominating the space sector, including novel spaceport services, will soon spill over to Europe, resulting in a *European New Space* dynamic. States need to be aware of the legal consequences of space activities taking place in, or rather, from their territory. Several legal consequences derive from the Outer Space Treaty which deals with the obligations of States related to private space activities. States bear international responsibility for space activities carried on by a non-governmental entity. States also have to authorise and continuously supervise private space activities.

2 The European New Space: Spaceports as the Driving Force

New Spaces describes "a global trend encompassing an emerging investment philosophy and a series of technological advancements leading to the development of a private space industry largely driven by commercial motivations."[2] *European New Space* signifies the current development of space industry on European territory. The global space economy grew by 6.7% on average per year between 2005 and 2017.[3] In 2019, the worldwide revenue of the global space industry amounted to U.S. $366 billion.[4] From an economic perspective, the engagement of private actors may generate noteworthy revenue. The growth in the global space sector has been influenced by rapid improvements in science and technology and an increasing demand for space-enabled services across a broad range of industries and countries.[5] Spaceports are part of this new trend, and private corporations are an important driving force. A spaceport can be understood as "spacecraft launching location".[6] Spaceports principally perform the same functions for spacecraft and rockets that traditional airports do for aircraft, namely facilitating landings and take-offs. In the United States, private corporations have already constructed, or plan to construct spaceports. Virgin Galactic's *Spaceport America* in New Mexico is the world's first purpose-built commercial spaceport, and Space X has announced similar plans for the

[2] European Investment Bank, *The Future of the European Space Sector* (2018) 3.

[3] Ibid.

[4] Satellite Industry Association, *Satellite Industry Report* (2019) 5.

[5] Deloitte Access Economics, *New Zealand Space Sector: its Value, Scope and Structure* (2019) 10.

[6] Stephen O 'Meara and Julias Dasch (eds), *A Dictionary of Space Exploration* (3rd edn, Oxford University Press 2018).

Starbase spaceport in South Texas.[7] Both companies plan to dedicate their spaceport for space tourism activities, especially under the auspice of reusable spacecraft.[8]

Europe is home to successful space launch programmes[9] and due to their legacy Europe is currently experiencing a strong entrepreneurial spirit dedicated to fostering private space launch activities. Commercial spaceports present an opportunity for the economic growth of remote and sparsely populated regions in Europe, and several European initiatives try to take this chance. In the United Kingdom, the commercial spaceport Cornwall at the Cornwall Airport Newquay purports to conduct the first launch in 2022.[10] Moreover, there are plans for constructing a spaceport for the vertical launch of spacecraft on the northern coast of Scotland.[11] In Italy, plans for suborbital space transportation are essentially connected to the development of new spaceports. The Italian approach aims to incorporate spaceport services in existing airports instead of constructing new spaceports from scratch.[12] Indeed one can draw a comparison between spaceports and airports as they have similar functions and provide similar services. One can even assume that spaceports will eventually have the same structures of ownership as airports do. The history of air transportation shows that the roots of civil air transportation lay in the military use of aircrafts during war and slowly shifted to the commercial exploitation of air transportation.[13] Over 20% of European airports are already privatised or are run as public–private partnerships. The largest European airports that account for the highest turnover of passengers are handled by privately owned airports.[14] Additionally, most publicly owned European airports are now run as corporatised entities and therefore, are treated like any other competitive business.[15] A similar business transformation will also expose spaceports to a competitive pressure where private actors will take the role as infrastructure providers. Besides relying on public funding or using public–private partnerships to provide public infrastructure,[16] spaceports will gradually become

---

[7] https://www.wired.com/story/spacex-reveals-texas-starport-plans/ accessed 6 August 2021.

[8] For an overview of national spaceports of other spacefaring nations see, Federal Aviation Administration, Office of Commercial Space Transportation (FAA/AST), *Commercial Space Data* https://www.faa.gov/data_research/commercial_space_data/ accessed 14 July 2021.

[9] See for example the French oversea spaceport Guiana Space Center supported by the European Space Agency, <https://www.arianespace.com/spaceport-facility/> accessed 15 July 2021.

[10] See the website of Spaceport Cornwall at <https://spaceportcornwall.com/> accessed 15 July 2021.

[11] Hamish Johnson, 'UK unveils plans for first spaceport' (2018) 31 Physics World 12.

[12] Francesco Santoro et al., 'Spaceport and Ground Segment assessment for enabling operations of suborbital transportation systems in the Italian territory' (2018) 152 Acta Astronautica 396.

[13] Peter Sand, Jeorge De Sousa Freitas and Geoffrey Pratt, 'A Historical Survey of International Air Law since 1944' (1960) 7 McGill Law Journal 128.

[14] Airports Council International, *The Ownership of Europe's Airports* (2010) 4.

[15] Ibid.

[16] The World Bank Group, *Public–Private Partnerships – Reference Guide Version 2.0* (2014) 14; Tare Brisibe, 'Prospect for the Arbitration of Disputes in Public – Private Space Projects' in Patricia M Sterns and Leslie I Tennen (eds), *Private Law, Public Law, Metalaw and Public Space Policy in Space* (Springer 2016) 53–57.

privately financed facilities. The goal lays in the expansion of the single European space market, which heralds the beginning of the *European New Space.*

3 The Role of Private Actors in International Space Law

The characterisation of an entity as private actor under the Outer Space Treaty[17] is relevant in two aspects: First, States are responsible for national activities in outer space carried out by non-governmental entities[18]; and second, States shall authorise and continuously supervise the activities of non-governmental entities in outer space. The status as "non-governmental entity" is precondition for several legal consequences but international space law does not define the term. Furthermore, the crucial question is whether spaceports amount to an "activity in outer space" if services are only provided on Earth.

3.1 International Responsibility of States for Spaceports of Non-governmental Entities

Pursuant to Article VI of the Outer Space Treaty, States shall.

> bear international responsibility for national activities in outer space (…) carried on by governmental agencies or by *non-governmental entities* (…).

The term "non-governmental entity" lacks a clear legal definition in the text of the Outer Space Treaty or in any follow-up legal document.[19] The Moon Agreement sets out a corresponding provision in Article 14, but also does not clarify the term "non-governmental entity".[20] Hence, it is necessary to interpret the term by referring to the rules of treaty interpretation and particularly to relevant rules of other

---

[17] Treaty on Principles Governing the Activities of States in the Exploration and Use of Outer Space, including the Moon and Other Celestial Bodies (adopted 19 December 1966, entered into force 10 October 1967) 610 UNTS 205 (Outer Space Treaty).

[18] Art. VI, Outer Space Treaty.

[19] Frans von der Dunk, *Private Enterprise and Public Interest in the European Spacescape* (International Institute of Air and Space Law, Leiden University 1998) 18.

[20] Art. 14, Agreement Governing the Activities of States on the Moon and Other Celestial Bodies (adopted 5 December 1979, entered into force 11 July 1984) 1363 UNTS 3 (Moon Agreement) reads: State Parties to this Agreement shall bear international responsibility for national activities on the Moon, whether such activities are carried on by governmental agencies or by non-governmental entities (…)." Stephan Hobe and Fabio Tronchetti, 'Article 14 Moon Agreement' in Stephan Hobe, Bernhard Schmidt-Tedd and Kai-Uwe Schrogl (eds), *Cologne Commentary on Space Law, Volume II* (Carl Heymanns Verlag 2009) 405.

fields of international law.[21] The Articles on Responsibility of States for Internationally Wrongful Acts stipulate certain general rules concerning the existence or consequences of the breach of an international obligation and attribution of these breaches, and may therefore be taken into consideration in order to determine the status of private actors.[22] The Outer Space Treaty introduces a specific rule of responsibility that differs from the general rules of state responsibility under the Articles on State Responsibility. The *lex specialis* responsibility regime of the Outer Space Treaty provides for the direct attribution of private conduct to the State.[23] In contrast, the secondary rules of the Articles on State Responsibility stipulate that the conduct of a private person is only attributable under certain, strict preconditions and in general assumes responsibility only for acts of State organs.[24] Even though the Articles on State Responsibility use different terms to describe the conduct in question, the object and purpose of the frameworks resembles each other. Both frameworks aim to ensure that international law is complied with[25] though Article VI Outer Space Treaty is dedicated specifically to space activities.[26] Both also differentiate between primary rules of liability and secondary rules of responsibility. Liability arises for injurious consequences of conduct not prohibited by international law, whereas responsibility requires the breach of an international obligation.[27] Moreover, international liability is based on the premise that absence of wrongfulness does not prejudge the question of compensation for damage caused by an act of a State.[28] These conceptual similarities between the Outer Space Treaty and the Articles on State Responsibility offer reason to refer to the rules of the Articles of State Responsibility to understand the term "non-governmental entity" better.

Article VI Outer Space Treaty as well as Articles 4-11 Articles on State Responsibility deal with the attribution of conduct to a State. While the Outer Space Treaty

---

[21] Art. 31 (3)(c) Vienna Convention on the Law of Treaties (adopted 23 May 1969, entered in to force 27 January 1980) 1115 UNTS 331 (VCLT).

[22] Articles on the International Responsibility of State for Internationally Wrongful Acts, UNGA Res 56/83, Annex (12 December 2001) (Articles on State Responsibility). James Crawford, 'The ILC's Articles on Responsibility of States for Internationally Wrongful Acts: A Retrospect' (2002) 96 American Journal of International Law 879. See also, Art. 55 Articles on State Responsibility.

[23] Michael Gerhard, 'Article VI OST' in Stephan Hobe, Bernhard Schmidt-Tedd and Kai-Uwe Schrogl (eds), *Cologne Commentary on Space Law, Volume I* (Carl Heymanns Verlag 2009) 110, 115.

[24] Art. 2, Art. 4–11, Articles on State Responsibility; International Law Commission, Articles on the International Responsibility of State for Internationally Wrongful Acts, with Commentaries, 'Report of the International Law Commission on the Work of its 53rd Session' (23 April–1 June and 2 July–10 August 2001) UN Doc A/56/10 (ILC Commentary) 38.

[25] Alain Pellet, 'The Definition of Responsibility in International Law' in James Crawford et al. (eds) *The Law of International Responsibility* (Oxford University Press 2010) 9.

[26] Michael Gerhard, 'Article VI OST' in Stephan Hobe, Bernhard Schmidt-Tedd and Kai-Uwe Schrogl (eds), *Cologne Commentary on Space Law, Volume I* (Carl Heymanns Verlag 2009) 104.

[27] James Crawford, 'State Responsibility' in *Max Planck Encyclopaedia of Public International Law* (2006 Oxford University Press) para. 4.

[28] Report of the International Law Commission on the Work of Its Thirty-Seventh Session, UN Doc. N40/10 (1985) 158.

regulates the attribution of national activities by "governmental agencies" or "non-governmental entities", the Articles on the Responsibility of States refer to the conduct of any "State organ",[29] "person or entity empowered to exercise elements of the governmental authority"[30] or the conduct of a "person or a group of persons".[31] In both frameworks, the international responsibility of a State can be invoked by another State for the conduct of either a public actor or a private actor. Even though the concrete attribution rules differ in their scope, the conceptual baseline is the same. A State is responsible for activities under its control. The Articles on State Responsibility stipulate different rules of attribution depending on the status of the actor. This differentiation of attribution is omitted in the Outer Space Treaty, where public and private actors are treated the same.[32]

The strict rules of attribution of private conduct and commercial activities to States under the Articles of State Responsibility reflects the initial separation between private international law and public international law. These contested public–private distinction concerned mostly aspects of State immunity and the exclusion from national court jurisdiction.[33] In contrast, the issue of private space activities focuses on the opposite concern, namely the categorization of a private conduct as public conduct. The topic of including private conduct in the domain of public international law was specifically discussed under the aspect of privatisation.[34] Privatisation effects the transfer of control from a public entity to a private one and hence raises the issue of attribution. Question of whether an act is private or public of private conduct and whether the status is either to be determined by national law or international law is ambivalently discussed under the Articles of State Responsibility.

Under the Articles on State Responsibility, the element of control differs based on the status of the actor.[35] Direct control is assumed for State organs, whereas for private actors effective control is required.[36] Under certain circumstances, a private spaceport operator's conduct may qualify as an act of State and therefore create responsibility. Since the conduct of State organs is considered an act of the State and thus creates attribution, it is important to determine what constitutes an organ in the sense of Article 4 Articles on State Responsibility. The term "State organs" has to be understood in its widest meaning, including de jure as well as de facto organs, such

[29] Art. 4, Articles on State Responsibility.

[30] Art. 7 and Art. 5, Articles on State Responsibility.

[31] Art. 8, Articles on State Responsibility.

[32] Frans von der Dunk, 'The Origins of Authorization: Article VI Outer Space Treaty' in Frans von der Dunk (ed), *National Space Legislation in Europe* (Brill 2011) 9.

[33] Peter-Tobias Stoll, 'State Immunity' in Max Planck Encylopedias of International Law (Oxford University Press 2011) paras. 13–15.

[34] Lorenzo Casini, 'Down the Rabbit-Hole: The Projection of the Public/Private Distinction beyond the State' (2014) 12 International Journal of Constitutional Law 402–428.

[35] Olivier de Foruville, 'Attribution of Conduct to the State: Private Individuals' in James Crawford et al. (eds), *The Law of International Responsibility* (Oxford University Press 2010) 257.

[36] *Case Concerning Military and Paramilitary Activities in and against Nicaragua (Nicaragua v. United States)* (Merits) (Judgment) [1986] ICJ Rep 14.

that all their conduct is attributable.[37] "State organs" pertain directly to the structure of the State.[38] The Articles of State Responsibility additionally establish the concept of entities empowered to exercise elements of governmental authority under Article 5 Articles on State Responsibility. The conduct of a person or entity that is not an organ of the State is nevertheless considered as an act of the State if the entity was empowered to exercise elements of governmental authority.

Article 5 Articles on State Responsibility uses the term "entity", which reflects a wide variety of bodies, including semi private entities or even private companies.[39] It reflects the need to include parastatal entities, which act instead of State organs, as well as situations of privatisation within the scope of State responsibility.[40] The only decisive criterion for attribution is whether the entity is empowered to exercise elements of governmental control.[41] Yet, it remains ambiguous whether "governmental authority" is determined by national law or public international law. The ILC Commentary on the Articles of State Responsibility elaborates:

> (…) the conduct of certain institutions performing public functions and exercising public powers (e.g. the police) is attributed to the State even if those institutions are regarded in internal law as autonomous and independent of the executive government.[42]

Additionally, the Commentary explains that the classification of the entity with reference to the ownership of its assets is of no relevance.[43] The interdependence of public international law and national laws is yet not clear. The ILC Commentary implies that public international law has its own concept of determining governmental functions, yet does not exclude national laws. It explains:

> Beyond a certain limit, what is regarded as "governmental" depends on the particular society, its history and traditions. Of particular importance will be not just the content of the powers, but the way they are conferred on an entity, the purposes for which they are to be exercised and the extent to which the entity is accountable to government for their exercise (…) The formulation of article 5 clearly limits it to entities which are empowered by internal law to exercise governmental authority.[44]

Multiple situations exists in which a categorisation as either exercising governmental authority or exercising strictly commercial activities proofs to be difficult. Entities may exercise regulatory authority, for example by concluding a contract with the State or by being granted a license, and yet national law does not define the

[37] Art. 4, Articles on State Responsibility.

[38] Djamchid Momtaz, 'Attribution of Conduct to State: State Organs and Entities Empowered to Exercise Elements of Governmental Authority' in James Crawford et al. (eds), *The Law of International Responsibility* (Oxford University Press Law 2010) 239.

[39] ILC Commentary 43.

[40] ILC Commentary 43.

[41] ILC Commentary 43.

[42] ILC Commentary 39.

[43] ILC Commentary 43. This is due to the fact, that a State may not evade responsibility by creating a private entity according to the criteria of a national legal system.

[44] ILC Commentary 43.

conduct as involving the exercise of public authority. Article 5 does not offer a satisfying solution for these cases.[45] One particular example is that of airline companies that exercise immigration control because the State authorised them to do so.[46] The immigration control is part of governmental authority and if the States has empowered the private airline company to conduct the control, then the State is also responsible for the airline companies' immigration control. For spaceports, certain conducts may qualify as exercising authority and therefore trigger the responsibility of the State. Yet it is important to notice that, under the Articles on State Responsibility, the State would only be responsible for spaceport services for which State powers have been granted, but not for other commercial activities.

Another field of law that deals with the distinction of public actors and private actors is international investment law.[47] In investment disputes, tribunals have to decide to which extent the determination of the entities conduct is based on national laws or international law. Tribunals try to solve this dilemma by applying a context-sensitive approach.[48] In the context of a breach of contract, the question arise whether a State is responsible for acts of a private entity that affects a foreign investor.[49] Moreover, it is crucial to decide whether an investor is "private" or "public", because different dispute settlement rules apply for the "private" investor and his "home State".[50] Several decisions dealt specifically with the question of attribution under the Article of State Responsibility.[51] In *EDF [Services] Ltd v Romania* the matter of concern was the arbitrarily taking of a concession to provide retail services at the Romanian airport Otopeni. The tribunal had to decide whether the actions of the airport operator were attributable to Romania. The tribunal focused on the rules of attribution pursuant to Article 5 Articles of State Responsibility and relied for deciding whether an entity was empowered to exercised governmental authority on two aspects. First, whether the internal law of the State concerned granted governmental authority; and second, whether the specific acts in question were an exercise of such delegated governmental authority[52] The tribunal found that, "(a) distinction

---

[45] ILC Commentary 43.

[46] Djamchid Momtaz, 'Attribution of Conduct to State: State Organs and Entities Empowered to Exercise Elements of Governmental Authority' in James Crawford et al. (eds), *The Law of International Responsibility* (Oxford University Press 2010) 244.

[47] Gus van Harten, 'The Public–Private Distinction in the International Arbitration of Individual Claims against the State' (2007) 56 International and Comparative Corporate Law Quarterly 171.

[48] Alex Mills, 'State Responsibility and Privatisation: Accommodating Private Conduct in a Public Framework (2021) <https://www.ejiltalk.org/state-responsibility-and-privatisation-accommodating-private-conduct-in-a-public-framework/ > accessed 9 August 2021.

[49] *SGS Société Générale de Surveillance S.A. v Pakistan* (Award), ICSID Case No. ARB/01/13 (2003); *SGS Société Générale de Surveillance S.A. v Philippines* (Decision on Jurisdiction), ICSID Case ARB/02/6 (2004); *Dede v Romania* (Award), ICSID Case No ARB/10/22 (2010).

[50] Rudolf Dolzer and Christoph Schreuer, *Principles of International Investment Law* (2nd edn, Oxford University Press 2012) 232–235.

[51] In *Toto Costruzioni Generali SpA v Lebanon* (Decision in Jurisdiction), ICSID Case No ARB/07/12 (2009) the tribunal was confronted with the claim of States that act by entities with a separate legal personality cannot be attributed to the State.

[52] *EDF (Services) Ltd v Romania* (Award), ICSID Case No ARB/05/13, IIC 392 (2009) para. 194.

has to be made between the legal regime of public property at the airport (such as runways, embarking or disembarking platforms or taxiways)", which is held and managed by the private entity and the "legal regime of the entity's private property which is a part of its own patrimony (such as all retail and other commercial spaces at the airport)." *In concreto*, the auctions of commercial spaces at the Otopeni Airport were not acts performed in the exercise of delegated governmental authority but were performed as purely commercial acts.[53] The tribunal therefore applied a functional approach to decide whether the acts of a private company can be attributed to the State. It decided on the bases of the particular case and the specific actions in question whether they may be attributed to the State. A similar functional approach could be applied to attribute responsibility of non-governmental entities in the sense of Article VI Outer Space Treaty. Governmental functions would render an entity as "governmental agency" whereas commercial functions would qualify an actor as "non-governmental entity". Especially for spaceports, a similar distinction of functions can be applied as in the case of *EDF [Services] Ltd v Romania.*

Yet, a dichotomy between the two responsibility regimes appears. The foremost purpose of the *lex specialis* responsibility rule of the Outer Space Treaty is to include any kind of private space activity irrespective of whether the activity is connected with State powers. Under the framework of the Outer Space Treaty, there is no need for governmental authority or effective control or posterior endorsement. The space-specific attribution rule is valid both for public as well as private actors.[54] In contrast, under the *lex generalis* of the Articles on State Responsibility, private or commercial activities are generally not attributable to the State.[55] The specific rule of direct attribution of private conduct of the Outer Space Treaty provides an exemption to the general rule of attribution under the Articles of State Responsibility. Any private conduct related to space activities is attributable to the State in contrast to the general rule that the only attributable conduct to the State is that of its organs of government, or of others who have acted under the direction, instigation or control of those organs.[56]

An indicator for a non-governmental entity could also be the ownership status of the entity concerned. The ownership status is also a factor relevant for attributing a conduct under Article 8 Articles of State Responsibility. Article 8 stipulates that the person or group of persons must be acting on the instruction of, or under the direction or control of, the State in carrying out the conduct the attribution of which is in question. Attribution under Article 8 for private conduct is set to be exceptional. Where the State uses its ownership interest in control of a corporation specifically in order to

[53] Ibid para 195.

[54] Sergio Marchisio, 'National Jurisdiction for Regulating Space Activities of Governmental and Non-Governmental Entities' (United Nations/Thailand Workshop on Space Law, Bangkok, 16–19 November 2010).

[55] ILC Commentary 43.

[56] ILC Commentary 38.

achieve a particular result, the conduct in question has been attributed to the State.[57] In *EDF [Services] Ltd v Romania*, the tribunal held that "(t)he fact that directions are given by the mandates to the members of the board of directors, a body that should decide in full autonomy in the company's interest, is indicative of the compelling nature of the Ministry's mandate system." The particular system of mandates issued by the State led to the direct influence of the State on the direction and control of the private entity's shareholders.[58] The fact that the State initially establishes a corporate entity is not a sufficient basis for the attribution to the State under the Articles on State Responsibility.[59] However, in combination with the factual influence of the State, the ownership status may serve as an indicator for attributing responsibility. For spaceports, the opposite case applies, namely a private entity conducts a commercial business without the involvement of the State. Yet, the ownership status could also be used for determining whether an actor qualifies as "governmental agency" or "non-governmental entity" by relying on the element of control.

3.2 *Ownership Structures of Spaceports as an Indicator for a Non-governmental Entity*

The traditional understanding of subjects of public international law creates a differentiation between States as the "traditional" subjects and international organisations and individuals as "partial" subjects of international law.[60] The Outer Space Treaty as being part of the broader branch of public international law endorses a State centric conception. Even though the Outer Space Treaty intends to regulate State powers, private actors have become involved in the process of creating and interpreting its provisions. Given that private actors are generally regulated on a national level, national law needs to be integrated in the public international law order.[61] The increased economic relations between States and private actors and on the establishment public–private partnerships requires a "multilevel economic law perspective",[62] that is also relevant for the realm of space activities. Public–private partnerships may be the main legal constellations for spaceport in the beginning, but privately financed spaceports will gradually expand their market share. The focus therefore needs to

[57] James Crawford, *The International Law Commission's Articles on State Responsibility- Introduction, Text and Commentaries* (Cambridge University Press 2002) 112–113.

[58] *EDF (Services) Ltd v Romania* (Award), ICSID Case No ARB/05/13, IIC 392 (2009) para. 208.

[59] ILC Commentary 48.

[60] Malcom N Shaw, *International Law* (8th edn, Cambridge University Press 2017) 142; Kate Parlett, *The Individual in the International Legal System* (Cambridge 2011) 3, 36.

[61] Joel Trachtmann, 'The International Economic Law Revolution' (1995) 17 University of Pennsylvania Journal of International Law 33.

[62] Ernst-Ulrich Petersmann, 'The Future of International Economic Law: A Research Agenda' (European University Institute, Department of Law Working Paper LAW 2010/06) 2.

shift from the subjects to the objects of international space law. For the application and interpretation of the Outer Space Treaty, a participant driven perspective is necessary in order to determine the term "non-governmental entity". The Outer Space Treaty is *lex specialis* and intends to cover any kind of space activity including commercial space activity. As explained above, the ILC Commentary on the Articles of State Responsibility indicates a certain interrelation of public international law with national laws. This interrelation is also relevant for interpreting the Outer Space Treaty.

Ownership can take several forms, such as public ownership, mixed public–private ownership, or private ownership.[63] The decisive criterion is the number of shares equalling control via voting rights.[64] Public ownership can be defined as "ownership of enterprises by the government, or by a government-controlled body."[65] A government-controlled body refers to an entity whose structures are according to national law and whose shares are wholly owned by public authorities.[66] Entities under public ownership are not within the scope of "non-governmental entity" when performing governmental functions. In contrast, where the spaceport is completely under private ownership, i.e., not owned by the government or by a government-controlled body, the entity qualifies as "non-governmental". Under a mixed public–private ownership, the spaceport is owned by a combination of public authorities and private investors. Where private actors hold shares that grant them controlling voting rights in a mixed public–private company, the company qualifies as a "non-governmental entity". National laws are therefore relevant for deciding the role an actor either by prescribing their status by referring to the ownership status. In this respect, Article 4.2 of the ILC Draft Articles states that "[a]n organ includes any person or entity which has that status in accordance with the *internal law* of the State." Yet, the ownership status as set forward in the respective national laws of the concerned State alone is not sufficient to decide the role of an actor. National law is irrelevant where a State tries to escape international responsibility by delegating its State functions to entities to which it gives separate legal personality under its national law.[67] The second aspect that has to be taken into consideration are the concrete functions exercised by the entity. Therefore, in order to determine the term

[63] See for example the ownership structures of airports; Airports Council International, *The Ownership of Europe´s Airports* (2010) 4.

[64] Alexander Dyck, *Ownership Structure, Legal Protection and Corporate Governance* (Annual World Bank Conference on Development Economics 2000) 28.

[65] Nigar Hashimzade, Gareth Myles, and John Black (eds), 'Public Ownership' in *A Dictionary of Economics* (5th edn, Oxford University Press 2017).

[66] "Public authority" are entrusted with executive or policy-making functions and the term is a synonym for the word "government." See UNCITRAL, 'Legislative Guide on Public–Private Partnerships' (2020) 6.

[67] Kaj Hober, 'State Responsibility and Attribution' in Peter Muchlinski et al. (eds), *International Investment Law* (Oxford University Press 2008) 582. Discussion on the attribution of companies´ conduct resolve around whether the "corporate veil" is a mere device for evasion, see *Barcelona Traction, Light and Power Company Limited (Belgium v. Spain)* (Merits) (Judgment) [1970] ICJ Rep 3 para. 56–58.

"non-governmental entity" one has to assess the relevant national laws in light of the actual functions of the entity. A functional approach would align the special rules of attribution of the Outer Space Treaty with other pertinent rules of international law.[68] However, in order to create legal certainty further legal development will have to focus on finding a definition of the term "non-governmental entity".

The Articles of State Responsibility provide significant importance for the legal consequences connected with responsibility. Since the Outer Space Treaty does not prescribe a specific legal consequence arising from responsibility, it is necessary to refer to the secondary norms of the Articles of State Responsibility.[69] Firstly, the responsible State has to ensure that the private actor ceases the wrongful conduct and pledge not to repeat it. Secondly, different forms of reparation can be claimed such as satisfaction, restitution and compensation.[70] Spaceports by private actors may not be attributable under the Articles of State Responsibility. Yet, through the specific direct attribution rules of the Outer Space Treaty, States are still responsible even for purely commercial space activities, and compensation claims arising out of private spaceports' conduct are dealt with under the Articles of State Responsibility.

3.3 Authorisation and Continuing Supervision of Spaceports

Qualification as a non-governmental entity is important not just for attributing responsibility and the duty to pay compensation. The second legal consequence connected with the status as a non-governmental entity is that States need to authorise and continue to supervise the space activity of non-governmental entities.

Article VI second sentence of the Outer Space Treaty reads:

> The activities of non-governmental entities in outer space (...) shall require authorization and continuing supervision by the appropriate State to the Treaty.

It is therefore important to determine whether an entity qualifies as non-governmental or governmental since different legal consequences relate to the status of the entity under the Outer Space Treaty. The duty to authorise and supervise is a corollary of the direct attribution rule. It is simply in the State's own interest to regulate commercial space activities, as they are responsible for private actors. Authorisation

---

[68] Similarly, under the rules of State immunity, entities only enjoy immunity if they "perform acts in the exercise of the sovereignty authority of the State", see Art. 2(1)(b)(iii) UN Convention on Jurisdictional Immunities of States and their Property (adopted 2 December 2004) A/RES/59/38. The ILC Report on the Draft Articles on Jurisdictional Immunities of States and Their Property (1991) UN Doc. A/46/10 at para. 14 states in this respect that: (...) an entity which is entitled to perform acts in the exercise of sovereign authority as well as acts or private nature, immunity may be invoked only in respect of the acts performed on the exercise of sovereign authority."

[69] Eric David, 'Primary and Secondary Rules' in James Crawford et al. (eds), *The Law of Responsibility* (Oxford University Press 2010) 27.

[70] Art. 36, Articles on State Responsibility.

and supervision mean that States should adopt, revise or amend regulatory frameworks for space activities of private actors, including rules related to spaceports.[71] However, Article VI does not prescribe how authorisation and supervision shall be ensured or be implemented in national laws. Authorisation and supervision ought to ascertain that space activities are carried out in a safe manner and minimise risks associated with the activity.[72] In practice, several States have established systems of authorisation under their national legislation for commercial or scientific activities, either by creating national space rules or by applying existing national rules[73] Three different methods of complying with the obligation under Article IV can be identified. First, the enactment of a comprehensive outer space law, that covers all possible activities; second, a sector-based approach and the enactment of laws for a specific sector (such as launch activities, telecommunications or remote sensing); and third, amending or revising already existing laws to cover new space activities.[74]

If one now assumes that the State has authorised the spaceport activity and has consequently created national laws that authorise the spaceport provider to exercise authority (as is also the case of immigration control at airports), a double role of the entity is created. Under the Outer Space Treaty, the obligation to authorise and continuously supervise applies only in relation to a non-governmental entity. Even if national laws may include regulations on governmental as well as non-governmental entities,[75] the concrete obligation of authorisation and supervision relates only to non-governmental entities. However, under the Articles on State Responsibility, the moment the State has authorised an entity to exercise authority, the conduct is no longer treated as private, but as an act of the State itself.

States will need to reflect their national regulations on their applicability to spaceports. They may either create new rules on spaceport or also amend existing laws so to apply to spaceports. This allows States to regulate space activities according to special national interests and still comply with their international responsibility.

---

[71] Recommendation on National Space Legislation Relevant for the Peaceful Exploration and Use of Outer Space, UNGA Res 68/74 (16 December 2013) UN Doc A/Res/68/74. See for example on the aspect of sustainability of space activities and national legislation: UNCOPUOS, 'Guidelines for the Long-term Sustainability of Outer Space Activities' UN Doc AC.105/2018/CRP.20 (27 June 2018) 5.

[72] Recommendation on National Space Legislation Relevant for the Peaceful Exploration and Use of Outer Space, UNGA Res 68/74 (16 December 2013) UN Doc A/Res/68/74.

[73] Irmgard Marboe and Gerhard F Hafner, 'National Authorisation Mechanism in Implementation of the UN Treaties' in Frans von der Dunk (ed), *National Space Legislation in Europe* (Brill 2011) 32.

[74] Recommendation on National Space Legislation Relevant for the Peaceful Exploration and Use of Outer Space, UNGA Res 68/74 (16 December 2013) UN Doc A/Res/68/74.

[75] See for example the §3 of the Austrian Outer Space Act of 2011 (Weltraumgesetz) BGBl. I Nr. 132/2011, which does not limit the authorisation on a specific actor but simply requires a space activity for which it has territorial or personal jurisdiction. It states: "Space activities require authorisation by the Minister for Transport, Innovation and Technology." Cordula Steinkogler, 'Austrian National Space Law' (2021) Oxford Research and Planetary Science 8–9.

Different conditions can be implemented for the authorisation of space activities.[76] The authorisation of spaceports may be made dependent on provisions for space debris mitigation, the protection of the environment or criteria relating to the experience, expertise and technical qualifications of the operator.[77] New Zealand is a good example of a space sector primarily driven by commercial activity.[78] Distinctive to the growth path of New Zealand's space sector is the presence of a major commercial launch company, Rocket Lab,[79] which established a launch base in 2016. The sudden emerge of space activities on New Zealand's territory was the catalyst for the establishment of the New Zealand Space Agency and to the enactment of the Outer Space and High-altitude Activities Act in 2017.[80] Until the development of legislation governing space activities, New Zealand referred to a contractual relationship with Rocket Lab in order to enable a safe and secure launch.[81] As can be seen in New Zealand's case, the establishment of a commercial launch base can take States by surprise and the development of legislation to comply with international obligations for space activities may take some time. States may see themselves suddenly in the position of being space-faring nations without proper preparation.

In Europe, the United Kingdom has already taken the first steps to react the ambition of commercial spaceports with the Space Industry Act of 2018,[82] which explicitly refers to spaceports. It includes the requirement of special spaceport licenses and range control licenses for the provision of range control services by commercial entities, including tracking, surveillance and boundary control. This approach is novel because in other countries, range control is typically provided by the State. UK therefore strongly aims at including commercial and private entities in the space industry. With the Space Industry Act of 2018 the United Kingdom also abides by its duties to authorise and supervise the private space activity under the Outer Space Treaty.

---

[76] See for example Recommendation 4 of the Recommendation on National Space Legislation Relevant for the Peaceful Exploration and Use of Outer Space, UNGA Res 68/74 (16 December 2013) UN Doc A/Res/68/74.

[77] Ibid.

[78] Frans von der Dunk, 'Kiwis in Space: New Zealand´s "Outer Space and High Altitude Activities Act"' *Proceedings of the 60th Colloquium on the Law of Outer Space* (International Astronautical Federation 2017) 453–467.

[79] Rocket Lab is a US company with a New Zealand subsidiary founded and runby New Zealand entrepreneur Peter Beck.

[80] Steven Freeland, Kirsty Hutchinson and Val Sim, 'How Technology Drives Space Law Down Under: The Australian and New Zealand Experience' (2018) 43 Air and Space Law 141–143. The text of the New Zealand Outer Space Act can be accessed at <https://www.legislation.govt.nz/act/public/2017/0029/latest/DLM6966275.html> .

[81] Ibid.

[82] The text of the UK Space Industry Act can be accessed at <https://www.legislation.gov.uk/ukpga/2018/5/contents/enacted/data.html> .

4 The Qualification of Spaceports as "Space Activity in Outer Space"

Spaceports services can develop like classical airport services.[83] Airports do not necessarily have their own aircraft and are not necessarily constructed for the sole purpose take-offs and landings for their own airline; they mostly provide the services necessary for operating flights of other airline companies but may also offer commercial retail services. Likewise, spaceports do not necessarily have to launch their own space objects but can simply offer the infrastructure for outer space companies to procure the launch. These ground-based spaceports with different launching methods raise the question of whether they qualify as an activity taking place *in* outer space; Article VI Outer Space Treaty is applicable only if they do qualify as such.

The Outer Space Treaty sets forth that States shall bear international responsibility for national activities in outer space. The literal meaning of when an activity takes place "in" outer space is ambiguous. There are two opposing views. According to the narrower approach, an activity which makes outer space accessible, explorable or usable qualifies as space activity.[84] The broader approach takes the position that even an activity taking place on earth qualifies as space activity, if it is predominately and intentionally directed at outer space.[85]According to the *telos* of the Outer Space Treaty, a State should be responsible for any activity related to outer space.[86] In order to include any space related activity, the provisions of Article VI should apply even where the activity only partly takes place in outer space. Every launching necessarily takes place partly on Earth (for the lift-off) and partly in outer space (after leaving air space).[87] A broad interpretation is also in line with Recommendation 1 of the UN General Assembly Resolution on National Space Legislation, which recommends that the scope of space activities shall include the launch of objects into and their return from outer space as well as the operation of a launch or re-entry site.[88] National laws seem to follow the UN General Assembly Recommendation and cover a broad scope of activities, such as the operation and guidance of space objects or

---

[83] Matthew A Bentley, *Spaceplanes-From Airport to Spaceport* (Springer 2009) 5.

[84] Hanneke Louise van Traa-Engelmann, *Commercial Utilization of Outer Space – Law and Practice* (Nijhoff 1993) 18.

[85] Bing Cheng, 'Revisited: International Responsibility, National Activities and the Appropriate State' (1998) 28 Journal of Space Law 7, 19; Frans von der Dunk, *Private Enterprise and Public Interest in the European 'Spacescape'* (International Institute of Air and Space Law, Leiden University 1998) 13.

[86] Michael Gerhard, 'Article VI OST' in Stephan Hobe, Bernhard Schmidt-Tedd and Kai-Uwe Schrogl (eds), *Cologne Commentary on Space Law, Volume I* (Carl Heymanns Verlag 2009) 108.

[87] For the delimitation of outer space from airspace see Stephan Hobe, 'Article I' in Stephan Hobe, Bernhard Schmidt-Tedd and Kai-Uwe Schrogl (eds), *Cologne Commentary on Space Law, Volume I* (Carl Heymanns Verlag 2009) 18.

[88] Recommendation on National Space Legislation Relevant for the Peaceful Exploration and Use of Outer Space, UNGA Res 68/74 (16 December 2013) UN Doc A/Res/68/74.

the operation of launch facilities.[89] Spaceports are nothing less than the "operation of a launch or re-entry site". Spaceports prepare and arrange the launch on Earth and provide range control service for the launching act. Reusable launch systems led to spacecraft capable of returning from orbit to Earth, so that the factor of re-entrance is particular valid for spaceports. Even if the spaceport is not directly involved in the mission of the space object in outer space itself, the launch activity has to be regarded as a whole. The act of launching or the re-entrance cannot be segregated from the launching site. The launching activity at the spaceport is directed at outer space and particularly makes outer space accessible for the space object. National laws with a broad material scope fulfil the obligations of Article VI Outer Space Treaty; also potential obligations of the State in its role as launching State derived from the Liability Convention and the Registration Convention are covered. A broad interpretation includes more easily new technological advancements.

A good example can be found in the space legislation of the United Kingdom. The UK Outer Space Act of 1986 is applicable to space activities pertaining to the launch of space objects by organisations or individuals established in the United Kingdom.[90] It sets forth all activities in Article I to which the Outer Space Act applies. *Inter alia* it lists the launching or procuring the launch of a space object as an activity that requires a license.[91] In 2018, the UK government enacted an Act specifically dedicated to commercial spaceflights and associated activities. The UK Space Industry Act of 2018 explicitly refers to the operation of a "spaceport". Under the UK Space Industry Act of 2018, commercial companies may apply for a license relating to launch from the UK and will replace the Outer Space Act of 1986.[92] This Space Industry Act of 2018 is of utmost relevance for the spaceport industry in the UK. Most notably, it introduces two different forms of spaceports, each requiring a license from a different authority. The Civil Aviation Authority regulates horizontal-launch spaceports, and the UK Space Agency regulates vertical-launch spaceports.[93] A horizontal-launch spaceport refers to "a spaceport with a runway—likely to be an adapted, existing aerodrome—suitable for launching spaceplanes

[89] Irmgard Marboe, Setsuko Aoki and Tane Brisibe, 'The 2013 Resolution on Recommendations on National Legislation relevant to the Peaceful Exploration and Use of Outer Space' in Stephan Hobe, Bernhard Schmidt-Tedd and Kai-Uwe Schrogl (eds), *Cologne Commentary on Space Law, Volume III* (Carl Heymanns Verlag 2009) 506–509.

[90] UK Space Agency, 'Spaceflight legislation and guidance' (29 July 2021) < Spaceflight legislation and guidance - GOV.UK (www.gov.uk)> accessed 2 August 2021.

[91] UK Space Agency, Department for Business, Energy & Industrial Strategy, Department for Transport, & Civil Aviation Authority, 'How we are promoting spaceflight from the UK' (29 July 2021) <How we are promoting and regulating spaceflight from the UK - GOV.UK (www.gov.uk)> accessed 2 August 2021.

[92] 'Understanding the Space Industry Act', 190208_Understanding_the_SIA_-_Final_For_Publication_-_Legal_Cleared_-_Initial_Publication.pdf (publishing.service.gov.uk).

[93] UK Space Agency, 'Applying for a future license under the Space Industry Act' (8 February 2019) <Applying for a future licence under the Space Industry Act - GOV.UK (www.gov.uk)> accessed 2 August 2021.

and carrier aircraft".[94] A vertical-launch spaceport refers to "a site—likely to be a new site—suitable for launching rockets." Subsection (2) specifies the other factors that regulators must consider, which also includes the international obligations of the UN space treaties. The different methods of launching requires to consider the delimitation of air space and outer space. In case of horizontal spaceports, the vehicle for launching is not a rocket but a spaceplane. A spaceplane, such as *SpaceShipOne*, is an evolution of an airplane into a vehicle capable of spaceflights.[95] Instead of launching vertically like a traditional rocket, spaceplanes take off from runways just like airplanes.

The two-fold approach of spaceports resurrects the problem of defining the scope of application of air law and space law, and the criteria for an "activity taking place in outer space" and not taking place in airspace. The relation of air law and space law appears constantly under different aspects.[96] The UNCOPUOS Legal Subcommittee comprehensibly analysed, on several occasions, the legal aspects of aerospace objects and the delimitation of outer space.[97] Whether the regime applicable to the flight of aerospace objects and space transport systems differ according to the object being located in airspace or outer space was a recurring issue.[98] States expressed divergent view with regard to the compatibility of the Chicago Convention[99] and the Outer Space Treaty, but the overall baseline was that international air law and international space law could apply to space transportation systems *mutatis mutandis*.[100] In 2020, the Secretariat of UNCOPUOS reported that member States still could not agree on a decision that would suit the interest of all States.[101] Therefore, the situation remains ambiguous and the categorisation of spaceports is left to the States' assessment.

The Chicago Convention applies to aircraft, which are defined pursuant to Annex 2 as "any machine that can derive support in the atmosphere from the reaction of the air (…)."[102] This is the case for spaceplanes and carrier aircrafts engaged in horizontal launches, to which a launch vehicle is attached and takes off during the flight. Accordingly, horizontal launch activities that include spaceplanes have to abide to Air Traffic Management rules set forward as Standards and Recommended Practices (SARPs) in the Annexes of the Chicago Convention. In contrast, vertical launches

---

[94] Louise Hughes, 'The Space Industry Act 2018' <PowerPoint Presentation (unoosa.org)> accessed 2 August 2021.

[95] Matthew A Bentley, *Spaceplanes-From Airport to Spaceport* (Springer 2009) 23.

[96] Suborbital flights are a similar issues that raises the question of mutual application of air law and space law. The International Civil Aviation Organization clarified in its statement to the Secretariat that the Chicago Convention also applies to suborbital vehicles, see UNCOPUOS, Note by the Secretariat A/AC.105/1039/Add.1 (2021) 8.

[97] UNCOPUOS Legal Subcommittee, A/AC.105/C.2/L.204 (Thirty-sixth session 1997); UNCOPUOS Report of the Secretariat A/AC.105/769 (2002) and A/AC.105/769/Add.1 (2020).

[98] UNCOPUOS Legal Subcommittee, A/AC.105/C.2/L.204 (Thirty-sixth session 1997) 5.

[99] Convention on International Civil Aviation (adopted 7 December 1944, entered into force 4 April 1947) 15 UNTS 295 (Chicago Convention).

[100] UNCOPUOS Legal Subcommittee, A/AC.105/C.2/L.204 (Thirty-sixth session 1997) 5 para 24.

[101] UNCOPUOS Report of the Secretariat A/AC.105/769 (2002) and A/AC.105/769/Add.1 (2020).

[102] Annex 2, Chicago Convention.

use rocket propulsion, and do not derive support in the atmosphere from the reaction with the air. Consequently, international air law as prescribed in the Chicago Convention is applicable during the launching phase from a horizontal launch spaceport. Whether these SARPs also apply to horizontal launches is doubtful because they do not meet the definition of an "aircraft".[103] New technological advancements further blur the line between vertical and horizontal launches. Some types of vertical-launch rocket engines depend on oxygen in the air while they are in the atmosphere and then switch to their stored oxygen tanks once they are out of the atmosphere. As a result, if a spaceport offers both methods of launching, horizontal and vertical, two different legal systems apply to the launching phase. The broader interpretation of a space activity would yet include both types of spaceports in the scope of the Outer Space Treaty. The application of air traffic management rules does not exclude the possibility of an object being covered both by air law as well as space law.

For instance, the German Aviation Code regulates that "space vehicles, rockets and similar flight objects are treated as aircraft as long as they are in air space."[104] This shows that a space activity can be subordinated to air traffic management rules, and still be covered by the Outer Space Treaty. In the absence of a single legal regime regulating aerospace objects and space transportation systems and the lack of legal clarity in the implementation of air law and space law, States may have to provide a categorisation in national laws. The need for national coordination between the aviation authority and the space authority is immanent. As the UK Space Industry Act shows, States already turn toward national legislation to provide a stable legal framework for space activities. National laws regulate the relationship between air space and outer space. It can be assumed that vertical spaceports classify as an "activity taking place in outer space" and hence are covered by Article VI Outer Space Treaty. The application of the Outer Space Treaty to horizontal spaceports is not as clear, yet it can be argued that horizontal spaceports are also covered by the broader definition of space activity taking place in outer space. Thus, under Article VI Outer Space Treaty, States are responsible for both, private actors' vertical- and horizontal-launch spaceports, and consequently have to create national rules to authorise and supervise these activities. The broad material scope of space activity would then cover both horizontal-launch spaceports as well as vertical-launch spaceports.

---

[103] Peter van Fenema, 'Legal Aspects of Launch Services and Space Transportation' in Frans von der Dunk (ed), *Handbook of Space Law* (Edward Elgar Publishing 2015) 410.

[104] §1 German Aviation Code (Luftverkehrsgesetz), 10 Mai 2007 (BGBL. I S. 698) last amended 5 July 2021 (BGBl. I S. 2287).

5 Liability for Damage Caused by Space Objects Launched from a Spaceport

The system of liability in international space law is based on States as the prime actors. Article VII Outer Space Treaty sets forth a general liability rule,[105] which is further specified by the Liability Convention.[106] The Liability Convention defines the term "launching States" and stipulates that the launching State is liable for damage caused by space objects to other States or their natural or juridical persons on Earth, in air space or in outer space. The term "launching State" is the decisive prerequisite for attributing liability. Accordingly, the "launching State" is defined as the State, which launches a space object, which procures the launching of a space object, or from whose territory or facility a space object is launched.[107] Besides being liable for damage as the first legal consequence, the launching State is also obliged to register its space objects in a national space registry and report the objects to the UN Secretary-General.[108]

The Liability Convention introduces two different liability regimes depending on where the damage occurs. Article II of the Liability Convention established absolute liability of the launching State for any damage caused by its space objects on the surface of Earth or to aircraft in flight. This particular protection of those suffering damage on Earth is unique in international law.[109] Absolute liability requires neither the element of fault nor proof of fault.[110] According to Article III, the launching State is also liable for damage caused by a space object elsewhere than on the surface of the Earth to the space object of another state only when at fault. The victim-oriented liability system is due to the fact that victims of damage on Earth shall be particular protected as they do not directly take part in the space activity.[111] Liability does not require a breach of an international obligation and absolute liability does not even need fault as prerequisite for compensation claims. Furthermore, Article XII of the Liability Convention stipulates that reparation shall restore the victim "to the condition which would have existed if the damage had not occurred." The Liability Convention therefor provides for the full and complete compensation of the victim.[112]

---

[105] Liability in the sense of Art. VIII Outer Space Treaty exists only in relation to State Parties of the Outer Space Treaty.

[106] Convention on International Liability for Damage Caused by Space Objects (adopted 29 November 1971, entered into force 1 September 1972) 961 UNTS 187 (Liability Convention).

[107] Art. I(c), Liability Convention; also Art. I(a), Registration Convention.

[108] Art. II-IV, Registration Convention.

[109] Malcolm N Shaw, *International Law* (8th edn, Cambridge University Press 2017) 888.

[110] Art. II, Liability Convention.

[111] Lesley J Smith, Armel Kerrest and Fabio Tronchetti 'Article II' in Stephan Hobe, Bernhard Schmidt-Tedd and Kai-Uwe Schrogl (eds), *Cologne Commentary on Space Law, Volume II* (Carl Heymanns Verlag 2013) 113–115.

[112] Frans von der Dunk, 'International Space Law' in Frans von der Dunk and Fabio Tronchetti (eds), *Handbook of Space Law* (Edward Elgar 2015) 84–86.

This liability system complements the general provisions of the Outer Space Treaty and encourage States to exercise control over space activities by establishing national space legislation.

The concept of the launching State is relevant for spaceport activities by private actors. The Liability Convention only refers to the "launching State" that launches or procures the launching, or from whose territory or facility the object is launched in its treaty text, which leads to a discussion on whether non-governmental activities are covered by the scope of the Liability Convention. A broad interpretation would include private launch facilities of and launch procurements by private actors.[113] One argument is that the Liability Convention has to be interpreted in the context of Article VI Outer Space Treaty.[114] This line of argumentation is also supported by the fact that Article III Liability Convention refers to "person for whom [the launching State] is responsible." The interaction between these two frameworks would lead to the inclusion of private actors in the liability regime. Especially the victim-oriented purpose of the Liability Convention deems it necessary to apply a broad interpretation. A narrow interpretation "would be counterproductive to the aims and goals" of the Convention.[115] Otherwise, States may evade their liability by privatizing their space activities.[116] The Liability Convention therefore includes all actors that conduct space activities in the sense of Article VI Outer Space Treaty. Consequently, private spaceports are covered by the scope of the Liability Convention. The State on whose territory the spaceport was constructed and consequently from whose territory objects are launched automatically qualifies as the launching State and is thus liable caused by space objects to other States or their natural or juridical persons on Earth, in air space or in outer space. Whether the spaceport is a private or a public launch site is irrelevant. The only common denominator is the "territory" from where the launch took place. For example, the spaceport Guiana Space Centre, from which Arianespace operates, is in the French overseas department of French Guyana. Consequently, France qualifies as the launching State for every launch from the Guiana Space Centre. Even if the spaceport is privatised, the State on whose territory the spaceport is operated, is liable for any damage caused

---

[113] Karl-Heinz Böckstiegel, 'The Terms "Appropriate State" and "Launching State" in the Space Treaties—Indicators of State Responsibility and Liability for State and Private Space Activities' in IISL (ed), *Proceedings of the International Institute of Space Law* (American Institute of Aeronautics and Astronautics 1992) 13–16.

[114] Lesley J Smith and Armel Kerrest, 'Article III' in Stephan Hobe, Bernhard Schmidt-Tedd and Kai-Uwe Schrogl (eds), *Cologne Commentary on Space Law, Volume II* (Carl Heymanns Verlag 2013) 134.

[115] Ibid 135.

[116] Armel Kerrest 'Liability for Damage Caused by Space Activities' in Marietta Benkö and Kai Uwe Schrögl (eds), *Space law: Current Problems and Perspectives for Future Regulation* (Eleven International Publishing 2005) 91–112.

by space objects launched from the spaceport.[117] It is also important to note that liability exists for damage during the launching phase and for attempted launches.[118]

The launching State has at least a financial interest in exercising control and regulatory powers on the spaceport. Financial compensation has no limit under the Liability Convention and States may face considerable compensation claims.[119] In this sense, the regime of liability has a preventive effect and States act more cautiously in their space endeavours due to their characterisation as ultra-hazardous activities. States may enter a contractual agreement dealing with the distribution of compensation, including hold harmless clauses and cross-waivers. They may also regulate who has to compensate a third party. However, this contractual construct is only possible between States.[120] The State can never pass liability to a private actor, even where the damage was caused solely by the private actor. If the State paid compensation to a third party, it may only seek recovery for the paid compensation. It is thus the launching State who bears the risk of the private actors' payment default or insolvency. It is in every launching State's interest that certain operational safety requirements are abided by spaceport operator. Damage cannot be averted to 100%, but with strict standards and a licensing system for spaceport activities, the risk of being exposed to compensation claims can be reduced. The Preamble of the Liability Convention observes "that notwithstanding the precautionary measures to be taken by States and international intergovernmental organizations involved in the launching of space objects, damage may on occasion be caused by such objects". Here is where the Outer Space Treaty and the Liability Convention complement each other. Especially where the launching State also qualifies as the responsible State in the sense of Article VII Outer Space Treaty, the interest of avoiding damage counterparts with national space laws prescribing safety standards for spaceport activities. National laws may also prescribe specific rules on insurance, debris mitigation, and market access that allows the launching State to minimise the risk connected with its status as launching State. Moreover, the absence of harmonised license requirements, launch information exchange, and air/space traffic control will put States in need of addressing these issues in their national laws as well.

---

[117] Lesley J Smith and Armel Kerrest, 'Article III' in Stephan Hobe, Bernhard Schmidt-Tedd and Kai-Uwe Schrogl (eds), *Cologne Commentary on Space Law, Volume II* (Carl Heymanns Verlag 2013) 132. The broad formulation of Art. III Liability Convention avoids a delimitation of outer space and air space, and covers any potential gap of Art.II Liability Convention.

[118] Art. I(b) Liability Convention which states that launching "includes attempted launches.".

[119] Lesley J Smith, Armel Kerrest and Fabio Tronchetti 'Article II' in Stephan Hobe, Bernhard Schmidt-Tedd and Kai-Uwe Schrogl (eds), *Cologne Commentary on Space Law, Volume II* (Carl Heymanns Verlag 2013) 124.

[120] Peter van Fenema, 'Legal Aspects of Launch Services and Space Transportation' in Frans von der Dunk (ed), *Handbook of Space Law* (Edward Elgar Publishing 2015) 397.

6 Conclusion

Private actors demand a broader involvement in the space industry and assume a role in the development of spaceports on European territory. Commercial companies are the driving force in the *European New Space* dynamic. The involvement of private actors in spaceport activities makes it necessary to reconsider central terminologies of international space law. Further legal development will have to focus on the definition of a "non-governmental entity" to create legal certainty. It may introduce a functional approach by referring to the ownership status of the spaceport, as well as to the functions exercised by the spaceport operator. States are well advised to prepare a legal framework on spaceports that consider the legal consequences connected with private spaceports. States bear international responsibility for private spaceports and non-compliance with international law is directly attributable to the State. Additionally, States have to authorise and continuously supervise private spaceport activities. The two possible categories of spaceports, namely horizontal-launch spaceports or vertical-launch spaceports, require careful consideration. This article suggests that horizontal- as well as vertical-launch spaceports are covered by the scope of the Outer Space Treaty. The authorisation and supervision of spaceports regulated by national space law may be a stimulating factor in the longstanding debate on the interrelation of air law and space law. Liability for damage is an additional factor that accompanies the issue of spaceports by private actors. The State from whose territory a space object was launched qualifies as the launching State, even when the launch was conducted by a private actor. Spaceports are covered by the scope of the Liability Convention and the launching States faces considerate compensation claims for damage caused by space objects, that were launched from the spaceport, to other States or their natural or juridical persons on Earth, in air space or in outer space. National laws on spaceports are advisable not only because States face responsibility and liability, but also because a stable legal framework on space activities attracts new investors. In the long run, a predictable legal environment will positively contribute to the advancement of spaceports in Europe and will foster the era of the *European New Space*.

Stephanie Stipsits works as a researcher and lecturer at the Department of European, International and Comparative Law at the University of Vienna. Stephanie studied Law at the University of Vienna and obtained an LL.M. in international economic law at Kobe University as a Monbukagakusho Scholar. She also worked at the Austrian Embassy in Tokyo, the Austrian Ministry of Defence and Sports, and at Oh-Ebashi LPD and Partners in Tokyo. Currently she is conducting research on State powers in areas beyond national jurisdiction and is writing her PhD thesis on Air Defence Identification Zones.

Upcoming UK Spaceports and the Future of Collaboration for Launches in Europe

Connor Hogan

Abstract The prospect of several new vertical and horizontal spaceports opening in the UK presents an invaluable opportunity for the European Space Agency (ESA) and for the commercial launch sector in Europe. However, cooperation on space activities has been complicated by the UK officially exiting the European Union in 2020, as although it remains a member of ESA and is set to collaborate on certain projects, it has had to leave most EU-funded space programs. This article will seek to explore the potential for future collaboration between European space partners and private actors on UK spaceports, in the context of Brexit and the UK's resultant third-party status within ESA. To do so, it will first give a brief account of the UK space sector within Europe, before exploring the impact of Brexit and the current regulatory environment for spaceports in the UK. From there, it will survey current proposals for UK spaceports. Finally, it will explore current and future areas of collaboration between the UK and the rest of Europe on UK-based spaceports, in light of political concerns.

1 Introduction

On 28 October 1971, the UK launched Prospero X-3.[1] A small scientific satellite, it was lofted to low Earth orbit (LEO) on a Black Arrow (R3) rocket, from Woomera Test Range, in South Australia.[2] To date, this has been the only British satellite ever

[1] Foust J, 'Five Decades after Black Arrow, a Reawakening UK Launch Industry Aims for Bullseye—SpaceNews' (*SpaceNews*, 8 August 2018). <https://spacenews.com/five-decades-after-black-arrow-a-reawakening-uk-launch-industry-aims-for-bullseye/> accessed 10 August 2021.

[2] TG Roberts, Spaceports of the World (Center for Strategic and International Studies 2019) 47.

C. Hogan (✉)
School of Politics and International Relations (SPIRe), University College Dublin (UCD), Dublin, Ireland
e-mail: connor.hogan@ucdconnect.ie

A. Froehlich (ed.), *Spaceports in Europe*, Studies in Space Policy 34,
https://doi.org/10.1007/978-3-030-88311-9_4

to be launched with a British rocket.[3] The UK's space program began in 1952,[4] and it began creating satellites in 1962, however until 1971, they had all been launched on American rockets.[5] The final Black Arrow flight was the first and only successful orbital launch by the UK, and following this, the British government (unique in this sense) ceased to pursue its own launch capabilities, citing cost and the existence of cheaper alternatives.[6]

Five decades after Black Arrow, the UK is seeking to rekindle its launch abilities,[7] with the aim of being the first country to launch satellites from Europe,[8] and of becoming a regional launch-hub for valuable polar and sun-synchronous orbits (SSO—also known as heliosynchronous orbits).[9] Such orbits are useful for imaging and weather satellites,[10] and are in high demand in the modern commercial satellite industry.[11] Geographically, the UK is advantageous for launching payloads to SSO and polar orbits, as it is at a high northern latitude (thus saving on fuel costs) and is surrounded by the sea (meaning launches can be conducted safely).[12] The UK already produces 40% of the world's small satellites,[13] and approximately a quarter of all telecommunications satellites.[14] A recent study found that fostering the ability for domestic launches could potentially generate up to £5.500.000.000 for the UK economy over the next decade.[15]

---

[3] Foust (n 2).

[4] Joanne Wheeler MBE and Vicky Jeong, 'The Law Reviews—The Space Law Review' (*The Law Reviews*, 17 December 2020) <https://thelawreviews.co.uk/title/the-space-law-review/united-kingdom > accessed 9 August 2021.

[5] Millard D, 'A Review of UK Space Activity and Historiography, 1957–2007' (2010) 66 Acta Astronautica 1291–95.

[6] Newman DCJ, 'The Space Industry Act 2018—Rediscovering the UK Sovereign Launch Capability' (2018) 2 ROOM: The Space Journal.

[7] Zakirov Z and others, 'The Market for a UK Launcher' (2018) 71 Journal of the British Interplanetary Society 399.

[8] Nick Flaherty, 'UK Aims to Be First European Country for Space Launches' (*eeNews Europe*, 24 May 2021) <https://www.eenewseurope.com/news/uk-aims-be-first-european-country-space-launches> accessed 8 August 2021.

[9] LaunchUK, 'A guide to the UK's commercial spaceports' (23 March 2021) <https://www.gov.uk/government/publications/brochure-a-guide-to-the-uks-commercial-spaceports> accessed 10 August 2021.

[10] Kopacz JR, Herschitz R and Roney J, 'Small Satellites an Overview and Assessment' (2020) 170 Acta Astronautica 93.

[11] Zakirov and others (n 8).

[12] Bill Read, 'The Magnificent Seven' (2021) 48 *Aerospace*.

[13] Gabriel Elefteriu, 'Britain's Industry-Led Space Policy "Model" Has Been a Resounding Success. But Can It Survive the Fierce Competition of the New Space Race?' (*Policy Exchange*, 30 May 2018) <https://policyexchange.org.uk/britains-industry-led-space-policy-model-has-been-a-resounding-success-but-can-it-survive-the-fierce-competition-of-the-new-space-race/> accessed 11 August 2021.

[14] ibid.

[15] Richardson M, 'New Study Reveals Multi-Billion Pound Potential of UK Space Industry—Aerospace Manufacturing' (24 March 2021) <https://www.aero-mag.com/new-study-reveals-multi-billion-pound-potential-of-uk-space-industry/> accessed 9 August 2021.

Moreover, there is a need for new spaceports. There is presently a launch 'bottleneck' for small satellites in the commercial sector,[16] and thus an increasing number of states are pursuing their own launch capabilities.[17] Indeed, although only twelve countries currently have active orbital launch facilities,[18] new spaceports are 'springing up'[19] in anticipation of 'a surge in the new space launch market and growing interest in suborbital flights for scientific experimentation.'[20]

The history of the British space industry is one built on integration and international alliances.[21] However, the UK's official withdrawal from the European Union (EU) in January 2020[22] and subsequent withdrawal from the single market and customs union in December 2020,[23] signals a possible challenge to this paradigm in the European context,[24] as the UK government seeks greater regulatory divergence from the EU. However as the European Space Agency (hereinafter ESA) is not an EU organisation, the UK's membership is in principle unaffected by Brexit.[25] Nevertheless, much of the political rhetoric in support of leaving the EU concerns state sovereignty, and a rejection of convergence with European regulations—an effort to 'take back control'[26]—which will likely inform public debate in all fields of European cooperation, including in space, for many years to come.

As Europe also seeks to develop its orbital launch capabilities, there is a perception of an emerging space race in the field of spaceports, wherein 'Brexit Britain'[27]

---

[16] Newman (n 7).

[17] Roberts (n 3).

[18] They are: Kazakhstan, Russia, the United States, China, French Guiana, India, Iran, Israel, Japan, New Zealand, and South and North Korea. See: Thomas G. Roberts, 'Spaceports of the World' (*Aerospace Security*, 4 January 2021) <https://aerospace.csis.org/data/spaceports-of-the-world/> accessed 11 August 2021.

[19] Quintana E, 'The New Space Age' (2017) 162 The RUSI Journal 91.

[20] ibid.

[21] Bleddyn Bowen, 'Leaving Spaceship Europe: British Space Policy after Brexit' (*British Politics and Policy at LSE*, 29 April 2018) <https://blogs.lse.ac.uk/politicsandpolicy/leaving-spaceship-europe-british-space-policy-after-brexit/> accessed 11 August 2021.

[22] Nigel Walker, *Brexit Timeline: Events Leading to the UK's Exit from the European Union* (House of Commons Library Briefing Paper SN 7960 CBP-7960, 2021) <https://researchbriefings.files.parliament.uk/documents/CBP-7960/CBP-7960.pdf> accessed 12 August 2021.

[23] ibid.

[24] Crawford H, 'The Future of the United Kingdom's Space Industry in a Post-Brexit World: Mitigating the Effects of Brexit on the UK'S Commercial Space Industry' (2021) 46 Air and Space Law.

[25] UK Department for Business, Energy & Industrial Strategy, 'UK Involvement in the EU Space Programme' (GOV.UK, 7 June 2021) <https://www.gov.uk/guidance/uk-involvement-in-the-eu-space-programme> accessed 8 August 2021.

[26] Menon A and Wager A, 'Taking Back Control: Sovereignty as Strategy in Brexit Politics' (2020) 8 Territory, Politics, Governance 279.

[27] Orbital Today, 'Will Brexit Britain Be Able To Conquer Space Alone & Create the UK Spaceports?' (*Orbital Today*, 2 April 2020) <https://orbitaltoday.com/2020/04/02/will-brexit-britain-be-able-to-conquer-space-alone-create-the-uk-spaceports/> accessed 5 August 2021.

wishes to outcompete European space partners by developing domestic launch facilities before those on the continent. However, the future relationship between European space actors and UK-based spaceports need not be as antagonistic as headlines may suggest. Indeed, LaunchUK, the UK's commercial spaceflight programme, has emphasised that they intend to work proactively with partners outside of the UK to ensure access to finance and marketing opportunities.[28] With this in mind, ESA and other partners in Europe could make use of the unique geographical and logistical advantages of UK launch sites in furthering their own space objectives.

This chapter will seek to explore the prospects of cooperation between the UK and the rest of Europe on UK-based spaceports, in the context of Brexit and surrounding political concerns. In the following section, the historical relationship between the UK, ESA and the EU will be discussed, followed by the impact of Brexit on UK involvement in the European space sector. From there, the current regulatory framework for spaceports in the UK will introduced, followed by a survey of the seven spaceports proposed for operation in the near future. Finally, the chapter will conclude with a discussion on current and future avenues of collaboration for European space partners on UK spaceports.

2 The UK Space Sector and Europe

The UK was one of the founding members of ESA, having ratified the Convention for the Establishment of a European Space Agency (hereinafter the ESA Convention) on 28 March 1978.[29] The Convention essentially aimed for the development of a Europe-wide space policy,[30] by 'elaborating and implementing activities and programmes in the space field'[31] and 'coordinating the European space programme and national programmes, and by integrating the latter progressively and as completely as possible into the European space programme'.[32] As there was no legal obstacle for Member States in developing their own national space programs in tandem with ESA, many of the largest European economies (e.g. France, Germany, Italy, Spain and indeed the UK) pursued (and continue to pursue) their own extensive national programs.[33] The UK founded the British National Space Centre (BNSC) in 1985 to coordinate domestic space policy, and enacted its first comprehensive, national space legislation

---

[28] LaunchUK (n 10) 2.

[29] Convention for the Establishment of a European Space Agency (ESA Convention) [1975] 1297 UNTS 161; UKTS 1981 No. 30; Cmnd. 8200; 14 ILM 864 <https://download.esa.int/docs/LEX-L/ESA-Ratification-acts/United_Kingdom_(28_Mar_1978).pdf> accessed 9 August 2021.

[30] Frans von der Dunk, Legal Challenges in the Context of the European Space Policy. in Thomas Hoerber and Sarah Lieberman (eds), A European Space Policy: Past Consolidation, Present Challenges and Future Perspectives (Routledge/Taylor & Francis Group 2019) 9.

[31] ESA Convention (n 30) Article IIb.

[32] ibid Article IIc.

[33] von der Dunk (n 31) 7.

in 1986 (discussed below).[34] In 2010, the BNSC was replaced with the UK Space Agency (UKSA).[35] As a signatory to the Outer Space Treaty (1967—hereinafter OST), the UK government is obliged to supervise its national space program to ensure it complies with the Treaty, as are all other parties.[36]

The UK's involvement within ESA has been deemed largely successful.[37] Indirect returns to the UK space sector from its contribution to the ESA budget (approximately £3.750.000 per annum) have at times reached an estimated £10 for every £1 invested.[38] Additionally, it was heavily involved in the development of flagship EU space missions, most notably the Galileo Project—to which it contributed 14% of the funding.[39] In 2019, the UKSA announced that it will invest £3.740.000 per year with ESA over the next five years.[40]

Dr Josef Aschbacher, Director General of ESA, speaking at a recent event organised by the UK's Policy Exchange think tank, emphasised the role that the UK continues to play in ESA missions, commenting that:

> International competition in space has never been at such a high point, and Europe needs to act to keep its standing in space. I expect the UK to be fully part of this exercise, of defining future European flagship programs, and to join forces to deliver the best space-based solutions for citizens.[41]

However, as will be discussed, the UK's future membership within ESA has become the subject of speculation considering the country's divergence from the rest of European policy and law, and the close relationship between ESA and the EU.[42,43]

In addition, in the so-called Second Space Age,[44] wherein space is becoming 'more accessible to smaller countries (…) and is increasingly commercial in its

---

[34] Wheeler and Jeong (n 6).

[35] C Housley and E Rough, 'The UK Space Industry' (22 April 2021). <https://researchbriefings.files.parliament.uk/documents/CBP-9202/CBP-9202.pdf> accessed 05 August 2021 8.

[36] Treaty on Principles Governing the Activities of States in the Exploration and Use of Outer Space, including the Moon and Other Celestial Bodies (opened for signature 27 January 1967, entered into force 10 October 1967) UNTS 205 Article VI.

[37] 'The Future of the European Space Agency and the UK's Role in It' (*Policy Exchange*, 10 May 2021) <https://policyexchange.org.uk/pxevents/the-future-of-the-european-space-agency-and-the-uks-role-in-it/> accessed 11 August 2021.

[38] ibid.

[39] Wells L, 'How Will Brexit Impact the Space Sector?' (*Open Access Government*, 11 September 2018) <https://www.openaccessgovernment.org/brexit-and-space/51943/> accessed 11 August 202.

[40] 'UK Invests in European Space Agency Programmes' (*GOV.UK*, 28 November 2019) <https://www.gov.uk/government/news/uk-invests-in-european-space-agency-programmes> accessed 12 August 2021.

[41] PolicyExchange (n 38).

[42] Crawford (n 25) 91.

[43] PolicyExchange (n 38).

[44] Harrison T and others, 'Escalation and Deterrence in the Second Space Age' (Center For Strategic & International Studies; Rowman & Littlefield 2017).

character rather than solely funded by public money',[45] a diverse range of private actors are now entering the industry. A recent government report noted a 'strong upward trend' in the number of private investments in UK space companies from 2013 to 2020 when compared to public investments,[46] and most planned departures from UK spaceports are with private companies.[47]

3 The Impact of Brexit

The UK commercial space sector depends on trade with the rest of Europe, with an estimated 53% of the industry's exports going to the continent, accounting for 19% of total income.[48,49] In 2018, 50% of satellite exports went to the single market, and construction costs were kept low by tariff-free trade with EU-base space manufacturers.[50] However, by removing itself from 'one of the biggest platforms for being heard in the vacuum of space,'[51] the UK potentially puts its position in the industry at risk, as it now has a comparatively smaller voice in the negotiation of regional space policy.[52,53] A recent survey of 1,218 UK-based space-related organisations showed that 'the most prevalent obstacle to commercial success' remains Brexit (55%), followed by Covid-19 (48%) and economic uncertainty (43%).[54] The EU's industrial space policy stipulates that all EU-funded ESA projects must be contracted to EU Member States within ESA,[55] and EU-funded projects do not need to provide industrial returns for ESA members outside of the EU.[56] Whereas countries which remain within the customs union and single market (such as Norway and Switzerland) may still have access to funds and services,[57] the UK has now left both, and so has been faced with the prospect of developing its own space projects (including spaceports) without EU support or funding. Moreover, as Member States construct

---

[45] Bowen (n 22) 1.

[46] Know.Space, 'Size & Health of the UK Space Industry 2020' (2021) Summary Report. <https://www.judiciary.uk/publications/administrative-court-judicial-review-guide-2018/> accessed 7 August 2021.

[47] LaunchUK (n 10).

[48] Crawford (n 25) 83.

[49] Know.Space (n 47) 4.

[50] Bowen (n 22).

[51] ibid 2.

[52] Crawford (n 25) 82.

[53] Bowen B, 'British Strategy And Outer Space—A Missing Link' (2018) 20 The British Journal of Politics and International Relations 335.

[54] Know.Space (n 47) 19.

[55] Commission (EC), 'EU Space Industrial Policy' (Communication) COM(2013) 108 final.

[56] Bowen (n 54) 336.

[57] Bowen (n 22) 2.

their own orbital launch facilities in the future, it is conceivable that a similar preferential regime could pertain for EU and ESA related launches, without comprehensive, pre-emptive negotiation.

Despite the aforementioned challenges, the UK space industry has actually expanded since the 2016 referendum: the latest government report found 'modest growth, a large increase in industry population (fuelled by new incorporations, signifying a healthy ecosystem), plus strong optimism for the next 3 years'.[58] The UKSA has been investing heavily in maintaining a competitive space sector, and aims to capture 10% of the global space market by 2030, up from 5.1% in the 2018–9 period.[59] The UK recently signed the Technology Safeguards Agreement with the US, which will enable firms to more easily import rocketry for use at UK spaceports.[60] Such international agreements will no doubt help the UK achieve its goals, but considering the aforementioned relative market share and geographical proximity to Europe, a necessary element in 'realizing its potential as a commercial satellite-launching hub'[61] will likely be thorough domestic regulation and continued cooperation with ESA and other European space partners.[62]

3.1 Exclusion from EU-Funded Programs

3.1.1 Galileo and EGNOS

As a result of Brexit, the UK has had to leave certain EU-funded space programs, most notably the Galileo Project and the European Geostationary Navigation Overlay Service (EGNOS).[63] Galileo is Europe's version of a global positioning system.[64] It will consist of thirty scientific satellites when completed,[65] and will cost an estimated €10 billion.[66] It is a project to which the UK heavily contributed—not just in terms

---

[58] Know.Space (n 47).

[59] ibid.

[60] Jonathan Amos, 'UK-US Sign Agreement on Space Launch Technology—BBC News' (*BBC News*, 17 June 2020) <https://www.bbc.co.uk/news/science-environment-53079844> accessed 11 August 2021.

[61] Crawford (n 25) 85.

[62] ibid 97.

[63] Payer M, 'UK Loses Galileo and EGNOS but Can Continue with Copernicus and ESA' (*SpaceWatch.Global*, 29 December 2020) <https://spacewatch.global/2020/12/uk-loses-galileo-and-egnos-but-can-continue-with-copernicus-and-esa/> accessed 12 August 2021.

[64] European Commission, 'Frequently asked questions—Galileo, the EU's satellite navigation programme' <https://ec.europa.eu/commission/presscorner/detail/de/MEMO_15_4711> accessed 11 August 2021.

[65] ibid.

[66] Wells (n 40).

of funding, but also in manufacturing.[67] The UK government expressed a desire to pursue its own version of Galileo upon leaving the project, however this was eventually abandoned.[68]

3.1.2 Copernicus

The UK has been able to negotiate continued participation in Copernicus,[69] 'the world's most ambitious program of space-based Earth observation'.[70] This is an EU program undertaken in coordination with ESA, which is tasked with the development and launching of the associated Sentinel satellites on behalf of the EU.[71] Elements of the Copernicus project require polar and sun-synchronous orbits, for which the UK is latitudinally suited. Both the Sentinel-3A and Sentinel-3B spacecrafts (which study ocean pollution and currents, the recession of the Earth's ice sheets and vegetation cover) make use of near-polar, high inclination orbits, which give optimal coverage of ice and snow parameters in high latitudes.[72] To achieve such orbits, they were launched from the Plesetsk Cosmodrome (roughly 800 kms north of Moscow[73]) in 2016 and 2018 respectively.[74,75] Similarly, the Sentinel-5 Precursor satellite (which studies air pollution) was launched from the Russian Far North.[76] The upcoming Sentinel-5 will be launched as part of the meteorological satellite MetOp-SG in early 2024, and will also require a polar orbit best accessed from higher latitudes.[77]

67 Amos J, 'UK Industry Bids Farewell to EU's Galileo System' *BBC News* (14 January 2021) <https://www.bbc.com/news/science-environment-55665537> accessed 12 August 2021.

68 Gallardo C, 'UK Scraps Brexit Alternative to EU's Galileo Satellite System' (*POLITICO*, 24 September 2020) <https://www.politico.eu/article/uk-scraps-plan-to-build-global-satellite-navigation-system-to-replace-galileo/> accessed 12 August 2021.

69 Department for Business, Energy & Industrial Strategy (n 26).

70 Clark S, 'European Environmental Observer Launched by Russian Rocket—Spaceflight Now' (25 April 2018) <https://spaceflightnow.com/2018/04/25/european-environmental-observer-launched-by-russian-rocket/> accessed 6 August 2021.

71 Clark (n 71).

72 Bergin C and Graham W, 'Russian Rokot Launches Sentinel-3A' (*NASASpaceFlight.com*, 16 February 2016) <https://www.nasaspaceflight.com/2016/02/russian-rokot-sentinel-3a-launch/> accessed 6 August 2021.

73 Pultarova T, 'Europe Launches Satellite to Map Air Pollution in Unprecedented Detail' (*Space.com*, 13 October 2017) <https://www.space.com/38457-europe-launches-sentinel-5p-air-pollution-satellite.html> accessed 6 August 2021.

74 Bergin and Graham (n 73).

75 Clark (n 71).

76 Pultarova (n 74).

77 'Metop Series | EUMETSAT' (*EUMETSAT*, 15 April 2020) <https://www.eumetsat.int/our-satellites/metop-series> accessed 6 August 2021.

3.2 ESA Membership

As discussed, ESA is a separate organisation from the EU,[78] and therefore the UK's membership is in theory unaffected by Brexit. Indeed, the UKSA cites continued membership of ESA as a key tenet of its development strategy.[79] Yet in the context of Brexit, a crucial aspect of British space development now rests in the independence of ESA from the EU.[80] However, the EU is a significant stakeholder in the formation and implementation of European-wide space policy, and works alongside ESA, other intergovernmental organisations and Member States to coordinate projects like Galileo and Copernicus, as well as to provide support through funding initiatives like the European Commission's Horizon 2020.[81] The EU is ESA's single largest procurer,[82] and approximately 26% of ESA funding comes directly from the European Commission, with much of the rest from Member States.[83] There was an initial proposal for ESA to become an agency of the EU by 2014,[84] however this did not materialise. Nevertheless, it has been argued that ESA is 'increasingly an implementing agency of EU space policy'.[85] From the UK government's perspective, further integration in this direction risks 'rendering ESA so "EU-like" that the UK's continued involvement becomes untenable.'[86]

Thus, in light of the UK's withdrawal from the EU, the question of collaboration between ESA and the UK on spaceports may become as much a diplomatic as a logistical one. The EU is also staking a claim as an independent space actor on the world stage, and is planning to help foster its own launch capabilities on the continent.[87] All of these factors represent potential challenges to the UK space sector, and will affect the degree of usage and cooperation on UK spaceports. In consideration of these challenges and potential future barriers to trade with the EU, the UK has invested heavily in its domestic launch abilities, and in efforts to create a legal and regulatory framework for spaceports.[88]

---

[78] 'ESA and the EU' (*ESA*) <https://www.esa.int/About_Us/Corporate_news/ESA_and_the_EU> accessed 12 August 2021.

[79] Department for Business, Energy & Industrial Strategy (n 26).

[80] Bowen (n 54) 335–6.

[81] European Commission, 'Horizon 2020' <https://ec.europa.eu/programmes/horizon2020/en/home> accessed 12 August 2021.

[82] Wells (n 40).

[83] PolicyExchange (n 38).

[84] ESA Communications 'Agenda 2011' (Summary Document) BR-268, September 2007. 25 <http://www.esa.int/esapub/br/br268/br268.pdf> accessed 11 August 2021.

[85] Bowen (n 22) 1.

[86] Crawford (n 25) 91.

[87] Tkatchova S, *Emerging Space Markets*, vol 35 (Microcosm Press and Springer 2018) 117–25.

[88] Maciel T, 'The British Space Explorer's Guide to Brexit' (*The National Space Centre*, 10 2018) <https://spacecentre.co.uk/blog-post/space-explorers-guide-to-brexit/> accessed 12 August 2021.

4 Current Regulatory Environment for UK Spaceports

4.1 The Space Industry Act 2018

The UK's increasing support for domestic space launches was demonstrated by the granting of Royal Assent to the Space Industry Act 2018 (hereinafter SIA). The SIA is a new legal framework which regulates inter alia licensing, liability and indemnity implications for UK-based space activities.[89] The UK government hailed the framework that the SIA lays out as 'one of the most modern in the world'[90] as it constitutes 'an ambitious attempt to re-establish independent launch capacity and a launch facility within the UK to compliment [sic] its burgeoning small satellite industry.'[91]

With the SIA as a regulatory foundation, the UK hopes to enable launches to commence from UK soil in 2022.[92] The legislation covers both orbital and suborbital activities,[93] not only for traditional rocket launches, but also air-launched vehicles,[94] space planes and high-altitude balloons.[95]

It has been speculated that the SIA came about largely as an attempt to 'future-proof'[96] the UK space industry against Brexit, and to take advantage of a growing space tourism and small satellite industry[97] 'with or without the ESA.'[98] It is the first major addition to UK space law since the Outer Space Act 1986 (hereinafter OSA), which was generally perceived as being 'overly burdensome on small companies' by requiring an expensive and rigorous licensing assessment process,[99] and overly punitive insurance requirements.[100] The result of adding the SIA is a 'twin track

---

[89] Smith LJ and Leishman RJM, 'Up, up and Away: An Update on the UK's Latest Plans for Space Activities' (2019) 44 Air & Space Law 1.

[90] Department for Transport and UK Space Agency, 'Space industry regulations consultations: summary of views received and government's response' (29 July 2020) <https://assets.publishing.service.gov.uk/government/uploads/system/uploads/attachment_data/file/968535/Space-Industry-Regulations-Consultations-summary-of-views-and-government-response-accessible.pdf> accessed 09 August 2021 4.

[91] Newman (n 7).

[92] LaunchUK (n 10).

[93] Newman (n 7).

[94] LaunchUK (n 10) 2.

[95] Flaherty (n 9).

[96] Newman (n 7) 1.

[97] ibid.

[98] Crawford (n 25) 95.

[99] Christopher J. Newman and Michael J. Listner, 'The Space Review: A Very British Coup: Lessons from the Draft UK Regulations for CubeSats' (*The Space Review*, 31 August 2015) <https://www.thespacereview.com/article/2816/1> accessed 9 August 2021.

[100] Newman (n 7) 2–3.

approach'[101] to UK space licensing, whereby the OSA concerns UK space activities occurring and launching abroad, and the SIA concerns UK-based launches and spaceport facilities.[102,103]

The text of the SIA has been described as 'skeletal'[104] in some regards, and has also been interpreted as a foundation on which to build, rather than a comprehensive regulatory framework.[105] Nevertheless, it does provide an initial legal basis for UK spaceports. Professor of Space Law and Policy at Northumbria University, Christopher Newman notes that:

> Indeed, to this end, the bulk of the SIA resembles a piece of planning legislation. There is provision for the development and operation of range control services, powers in relation to land that will be used for spaceports, safety and security at spaceports and provisions for ensuring that licenses will not be granted unless the operator has submitted an environmental impact assessment.[106]

Under the SIA, ground operations are to be regulated by the Health and Safety Executive, orbital launches by the UKSA, and suborbital activities by the Civil Aviation Authority (CAA).[107] Authors Lonsdale and Phillips[108] note that the SIA and subsequent legislation has provided a framework for environmentally safe space launches, although as discussed below, this remains a consistent obstacle to public assent for UK spaceports.

4.2 *The Space Industry Regulations 2021*

In 2020, the UK government ran two consultations with the public and industry specialists on draft secondary legislation for the implementation of the SIA.[109] In March, the results of these consultations were published, and in May of 2021, the government laid new regulations before Parliament, which provide specifically for licensing and regulation of UK spaceports: the Space Industry Regulations 2021.[110,111] Additional, concurrent space regulations came in the form of the

[101] ibid 4.

[102] ibid 4–5.

[103] Wheeler and Jeong (n 6).

[104] Newman (n 7).

[105] Crawford (n 25).

[106] Newman (n 7) 5.

[107] ibid.

[108] Lonsdale J-A and Phillips C, 'Space Launches and the UK Marine Environment' (2021) 129 Marine Policy 104,479.

[109] Department for Transport and UK Space Agency (n 91).

[110] UK Government, 'British Spaceflight to Become Reality as Government Provides Launchpad for Spaceports' (*GOV.UK*, 24 May 2021) <https://www.gov.uk/government/news/british-spaceflight-to-become-reality-as-government-provides-launchpad-for-spaceports> accessed 5 August 2021.

[111] Space Industry Regulations 2021 Order 2021, SI 2021/792.

Space Industry (Appeals) Regulations 2021, the Spaceflight Activities (Investigation of Spaceflight Accidents) Regulations 2021 and the Contracting Out (Functions in Relation to Space) Order 2021, all of which came into force on 29th July 2021.[112] This package of statutory instruments now regulates the licensing of launch, return, orbital, spaceport and range control activities on UK soil,[113] and was developed in collaboration with UKSA and the UK Civil Aviation Authority.[114] As a result of these instruments, UK operators can now apply for a spaceport licence.[115] Spaceport licences are a requirement for the operation of a spaceport, which is defined as 'a site from which spacecraft or carrier aircraft are launched or (as the case may be) are to be launched'[116] or 'a site at which controlled and planned landings of spacecraft take place or (as the case may be) are to take place.'[117] The scope of activities covered under the licence include the provision of a safe ground environment for launches, maintaining public safety in the vicinity of the spaceport, the regulation of propellant transportation and storage, and on-site engine testing.[118]

5 Upcoming UK Spaceports

There are currently seven proposed spaceport sites in the UK, all of which are in the process of applying for operating licences and seeking out customers.[119,120] LaunchUK is working with a range of partners to establish vertical and horizontal launch capabilities for small satellites.[121,122] Additionally, the UK government is providing grants totalling almost £40.000.000 to help support new spaceports.[123]

Scotland is the best place in the UK to reach high-demand polar orbits,[124] as it is furthest north, and is surrounded by the Atlantic Ocean to the west, and the North Sea to the north and east.[125] Of the seven proposed spaceports, five are in

112 UK Space Agency and others, 'How We Are Promoting Spaceflight from the UK' (*GOV.UK*) <https://www.gov.uk/guidance/how-we-are-promoting-and-regulating-spaceflight-from-the-uk> accessed 10 August 2021.

113 Department for Transport and UK Space Agency (n 91).

114 UK Government (n 111).

115 ibid.

116 Space Industry Act 2018 Sect. 3(2)(a).

117 Space Industry Act 2018 Sect. 3(2)(b).

118 Wheeler and Jeong (n 6).

119 Read (n 13).

120 LaunchUK (n 10).

121 Housley and Rough (n 36) 25–7.

122 UK Space Agency and others (n 113).

123 ibid.

124 ibid.

125 'Why Is Scotland a Prime Rocket Launch Site?' *BBC News* (19 November 2020) <https://www.bbc.com/news/uk-scotland-48119118> accessed 10 August 2021.

Scotland. Matjaz Vidmar of the University of Edinburgh has described Scotland as the emerging 'Space Glen'[126] of a global New Space[127] economy, in reference to the development of a regional ecosystem of small-to-medium sized space enterprises there. In 2020, the Scottish Space Leadership Council formed the UK Spaceport Alliance—a consortium of all current potential spaceport operators based in Scotland.[128] Though the UK Spaceport Alliance currently only includes the Scottish spaceports, talks with the other UK-based sites are at 'an advanced stage' with the aim of creating a coordinated UK-wide approach to spaceport usage and planning.[129] Speaking on the establishment of the Alliance, Frank Strang, CEO of the SaxaVord Spaceport in Shetland, commented that:

> Recent speculation about a 'Space Race' between UK launch sites has got it wrong. The race is not within the UK, it is between the UK and the rest of Europe and we recognise that we have to work together and collaboratively to establish the UK as a leader not only in Launch, but across the Space sector as a whole.[130]

The above sentiment echoes the perception of spaceports as a new avenue of competition between the UK and the rest of Europe. It is crucial in this regard that policymakers act pre-emptively to ensure the highest degree of mutually beneficial cooperation between both space sectors in the field of launch facilities.

5.1 Vertical Spaceports

5.1.1 SaxaVord Spaceport

In January 2021, plans for a new vertical spaceport were submitted to the Shetland Islands Council by town planning consultants Farningham Planning Ltd, on behalf of Shetland Space Centre[131] (recently renamed to SaxaVord Spaceport[132]).

---

[126] Vidmar M, 'New Space and Innovation Policy: Scotland's Emerging "Space Glen"' (2020) 8 New Space 31.

[127] 'New Space' is a broad term used to describe a number of trends in global space policy, chief among them the commercialisation of space, the entry of a multiplicity of private actors, and increased cost-effectiveness. See: Davidian K, 'Definition of New Space' (2020) 8 NewSpace 53–5.

[128] 'UK Spaceport Alliance' (*Scottish Space Leadership Council*) <http://scottishspace.org/spaceport-alliance/> accessed 10 August 2021.

[129] ibid.

[130] ibid.

[131] 'Prospect of Major Boost to Shetland Economy as Space Launch Site Plans Lodged | Shetland Space Centre' (*Shetland Space Centre*, 18 January 2021) <https://shetlandspacecentre.com/general/prospect-of-major-boost-to-shetland-economy-as-space-launch-site-plans-lodged/> accessed 11 August 2021.

[132] 'Name Change and Rebrand for Our Rocket Site' (*Shetland Space Centre*, 30 June 2021) <https://shetlandspacecentre.com/general/name-change-and-rebrand-for-our-rocket-site/> accessed 11 August 2021.

The proposed spaceport is to be located on the Lamba Ness Peninsula, on the north-coast of Unst, in the Shetland Islands.[133] This is the UK's most northerly island,[134] and one of the highest latitudes in Europe,[135] making it ideal for efficiently launching small satellite payloads into polar and sun-synchronous orbits.[136] The site will enable the launching of satellites weighing up to 600 kg,[137] and is set to feature three launch pads, a satellite tracking facility, hangar and integration facilities, a range control centre and a fuel storage facility.[138] The spaceport is predicted to generate around 600 jobs in Scotland by 2024.[139]

The isolation of Unst means that two or three stage rockets can launch relatively safely, with the attendant debris falling into the North Sea.[140] The UK government is negotiating with Norway, the Faroe Islands and Iceland on agreements to allow stages to fall in their maritime jurisdictions.[141]

The site is the home of the UKSA's Vertical Launch Pathfinder Programme,[142] which is being undertaken by Lockheed Martin in conjunction with the UK government. Lockheed Martin does not have a small launch vehicle that is compatible with the spaceport however,[143] and so in February the company awarded a contract for the first satellite launch to ABL Space Systems, which is scheduled to launch the UK Pathfinder mission on its RS1 micro-launcher from SaxaVord Spaceport in 2022—potentially the first vertical satellite launch from UK soil.[144] Once in orbit, the RS1 will deploy a payload of CubeSats, built by the US company Moog in the UK.[145]

The SaxaVord spaceport is scheduled to have a relatively high launch capacity: it is hoped to have about 30 per year by 2024.[146] In addition to Lockheed Martin and ABL Space Systems, the spaceport announced in June 2020 that C6 Launch

[133] Shetland Space Centre (n 132).

[134] 'SaxaVord Spaceport Plans Take Step Nearer Launch' *BBC News* (22 October 2020) <https://www.bbc.com/news/uk-scotland-north-east-orkney-shetland-54634943> accessed 10 August 2021.

[135] LaunchUK (n 10).

[136] Read (n 13).

[137] Emma Thorpe, 'The Race Is on as the UK Goes Ahead with the Approval of the SaxaVord Spaceport' (*Orbital Today*, 2 November 2020) <https://orbitaltoday.com/2020/11/02/the-race-is-on-as-the-uk-goes-ahead-with-the-approval-of-the-shetland-space-centre/> accessed 10 August 2021.

[138] Read (n 13).

[139] BBC News (n 135).

[140] Read (n 13).

[141] ibid.

[142] Shetland Space Centre (n 133).

[143] 'Lockheed Martin Selects ABL Space Systems Rocket to Power First UK Vertical Satellite Launch' (*Lockheed Martin*, 8 February 2021) <https://news.lockheedmartin.com/news-releases?item=129013> accessed 10 August 2021.

[144] 'Lockheed Martin Selects ABL Space Systems Rocket to Power First UK Vertical Satellite Launch' (*Media—Lockheed Martin*, 8 February 2021) <https://news.lockheedmartin.com/news-releases?item=129013> accessed 10 August 2021.

[145] ibid.

[146] BBC News (n 135).

Systems, a Canadian small launch developer, has outlined plans to launch from the site.[147] In February 2021, the German rocket manufacturer HyImpulse Technologies announced plans to begin launching suborbital sounding rockets from SaxaVord by the end of the year, and aims to begin orbital launches from the spaceport in 2023.[148] The co-CEO of the company, Christian Schmierer, noted that they have 'signed letters of intent with several potential customers to take their payloads into orbit.'[149] In May, the company began engine testing in Shetland.[150] Paul Riddel, Head of Comms at the SaxaVord Spaceport, recently confirmed that they were in talks with additional customers, but was unable to specify prior to formal agreements.[151]

There has been some opposition to the spaceport, however, as it is located in an area of historical importance. In March 2021, Historic Environment Scotland (HES), an environmental statutory agency, refused consent for the development on the grounds that it would destroy 'a scheduled monument of national importance' (a former radar base located on the site).[152] As of August 2021, the dispute remains unresolved.

5.1.2 Space Hub Sutherland (SHS)

In August 2020, planning permission was granted to build a spaceport on the A' Mhòine Peninsula, in Sutherland, Scotland.[153] The development of the site is being supported by the UKSA and the Highlands and Islands Enterprise development agency (HIE),[154] who funded grants of £2.500.000[155] and £8.500.000 respectively.[156] When operational, Space Hub Sutherland will serve as the region's 'compact vertical

---

[147] Jeff Foust, 'Lockheed Martin Shifts U.K. Launch Site—SpaceNews' (22 October 2020) <https://spacenews.com/lockheed-martin-shifts-u-k-launch-site/> accessed 10 August 2021.

[148] 'German Rocket Maker Plans to Launch from Unst | SaxaVord Spaceport' (*The SaxaVord Spaceport*, 2 February 2021) <https://shetlandspacecentre.com/general/german-rocket-maker-plans-to-launch-from-unst/> accessed 10 August 2021.

[149] ibid.

[150] 'Successful Engine Tests for German Rocket Maker | SaxaVord Spaceport' (*The SaxaVord Spaceport*, 19 May 2021) <https://shetlandspacecentre.com/general/successful-engine-tests-for-german-rocket-maker/> accessed 10 August 2021.

[151] Read (n 13).

[152] Chris Cope, 'Historic Environment Agency Rejects Space Centre Plan to Build at Former Radar Base' *Shetland News* (30 March 2021) <https://www.shetnews.co.uk/2021/03/30/historic-environment-agency-turns-down-space-centre-plan-to-build-at-former-radar-base/> accessed 10 August 2021.

[153] LaunchUK (n 10) 10.

[154] Williamson M, 'Is the UK "Go for Launch"?: A Key Component in the UK's Ambition to Become a Powerhouse in the Space Industry Is to Have Indigenous Launch Capabilities' (2021) 16 *Engineering Technology* 32.

[155] BBC News (n 126).

[156] Read (n 13).

launch site',[157] capable of delivering small satellites (less than 500 kg[158]) to polar and sun-synchronous orbits,[159] and will be the only vertical launch facility on the UK mainland.[160] It will consist of a single launch pad, operations control centre and assembly building.[161] The spaceport is expected to generate around 40 jobs on-site, and a further 200 in the wider region.[162] The first launch is planned for 2022.[163]

SHS will be owned and operated by the HIE,[164] and is expected to be used exclusively to launch rockets for the UK-based Orbital Launch Express (Orbex) company.[165] A UK government contract with Lockheed Martin was originally linked to the Sutherland site, but was later moved to the SaxaVord spaceport[166] in order to avoid conflict with Orbex[167] and a backlog of launch dates,[168] as the local council has indicated that launches from Sutherland will be limited to 12 per year, to mitigate the amount of metal and plastic debris falling into the sea during launches.[169] Despite this, the HIE Project Director, Roy Kirk, indicated in early 2021 that they were in 'in discussions with a number of customers who are actively considering using [SHS].'[170]

The site has also faced some opposition from local groups on environmental grounds, and there is currently a legal challenge being brought by a local billionaire couple, Anders and Anne Holch Povlsen, who own land near the proposed site.[171] In early 2021, Anders Holch Povlsen announced that he had invested over £1.400.000 in the alternative spaceport in Shetland.[172]

---

[157] LaunchUK (n 10) 10.

[158] ibid.

[159] ibid.

[160] Read (n 13).

[161] LaunchUK (n 10).

[162] ibid.

[163] Williamson (n 155) 33.

[164] Read (n 13).

[165] Williamson (n 155).

[166] Foust (n 148).

[167] Williamson (n 155).

[168] Thorpe (n 138).

[169] Victoria Brenan, 'Scotland's Richest Man to Mount Legal Challenge over Spaceport' (*Herald-Scotland*, 19 January 2021) <https://www.heraldscotland.com/news/19020448.scotlands-richest-man-mount-legal-challenge-spaceport/> accessed 10 August 2021.

[170] Read (n 13).

[171] 'Billionaire Landowners Oppose Highlands Space Port Decision' *BBC News* (6 November 2020) <https://www.bbc.com/news/uk-scotland-highlands-islands-54825109> accessed 10 August 2021.

[172] Brenan (n 170).

5.1.3 Spaceport 1 (SP-1)

The third proposed vertical spaceport in Scotland would be located in Scolpaig on the north-west coast of North Uist, in the Outer Hebrides.[173] In June 2019 the Western Isles local council invested around £1.000.000 to purchase the land needed to build the spaceport.[174] The project is being coordinated by the HIE, the Commercial Space Technologies consultancy, and QinetiQ, which operates the nearby Ministry of Defence Hebrides Rocket Range.[175,176] The operators aim to facilitate suborbital launches of <100 kg by the end of the year, and orbital launches of <350 kg by 2023.[177] Much like Shetland, the isolated location means that Spaceport 1 will be able to facilitate three stage launch vehicles and provide sun-synchronous and polar orbits.[178]

The project has also come up against strong objections on environmental grounds, and progress has slowed since 2019.[179,180]

5.2 *Horizontal Spaceports*

Space laws have hitherto been predicated chiefly upon vertical rocket launches.[181] However the UK has also invested heavily in the development of horizontal launch and suborbital spacecraft, and has explicitly regulated for such activities through the SIA.[182]

5.2.1 Spaceport Cornwall

Spaceport Cornwall is a joint project supported by Cornwall Council, Cornwall and Isles of Scilly Local Enterprise Partnership, Goonhilly Earth Station and commercial launch operator Virgin Orbit,[183] and will be located near the town of Newquay.[184]

---

173 LaunchUK (n 10).

174 Read (n 13).

175 'Western Isles in Race to Open UK's First Spaceport' *BBC News* (11 June 2019) <https://www.bbc.com/news/uk-scotland-highlands-islands-48593633> accessed 10 August 2021.

176 Read (n 13).

177 LaunchUK (n 10) 18–19.

178 ibid.

179 Williamson (n 155) 32.

180 'Hundreds of Objections to North Uist Spaceport' *BBC News* (19 September 2019) <https://www.bbc.com/news/uk-scotland-highlands-islands-49753441> accessed 10 August 2021.

181 Newman (n 7) 4.

182 ibid.

183 LaunchUK (n 10) 7.

184 ibid.

The site boasts a 2.744 m long runway and has direct access to the sea.[185] When operational, it will have the capacity to launch up to 300 kg to LEO, with the first launch planned for 2022.[186] In January 2021, Virgin Orbit successfully launched 10 NASA CubeSats to LEO using a rocket launched from a horizontal craft.[187] The launch marked a 'positive juncture' for the UK's launch ambitions, as the launch technology has now been proven to work.[188] Virgin Orbit will be Spaceport Cornwall's primary initial customer, and hopes to use its horizontally-launched satellite system for a variety of customers.[189]

In addition to Virgin Orbit, interim head of Spaceport Cornwall, Mellissa Thorpe recently stated that the spaceport will be a 'multi-user site' and confirmed that they are in talks with 'numerous launch service operators at varying stages of technology readiness' to use the spaceport for future launches.[190]

5.2.2 Spaceport Snowdonia

Another proposed spaceport site is in Llanbedr, Gwynedd, Wales. The site is a former military base, which is owned by the Welsh government but has been leased to Snowdonia Aerospace—a regional operator.[191] Similarly to the site in North Uist, progress on Spaceport Snowdonia has been comparatively slow.[192] Nevertheless, Snowdonia Aerospace hopes to conduct the first suborbital flights in 2022–25,[193] as well as sun-synchronous and polar orbital launches for payloads of up to 200 kg.[194] A successful balloon test was conducted by the English company B2Space in 2020, as a feasibility study into the use of atmospheric balloons for micro-satellite launches.[195] In 2019, the UK government announced £500.000 in funding for the spaceport, and the Welsh government has invested £135.000 to test unmanned and electric aircraft and space planes.[196] The UKSA invested a further £86.000 to research satellite-enabled drone technology at the site.[197]

---

[185] ibid.

[186] ibid.

[187] Shead S, 'Richard Branson's Virgin Orbit Successfully Blasts Satellites into Space after Failed First Attempt' (*CNBC*, 18 January 2021) <https://www.cnbc.com/2021/01/18/richard-bransons-virgin-orbit-blasts-satellites-into-space-from-747.html> accessed 10 August 2021.

[188] Williamson (n 155).

[189] Read (n 13).

[190] ibid.

[191] ibid.

[192] Williamson (n 155).

[193] Read (n 13).

[194] LaunchUK (n 10) 12–3.

[195] Kate Rimmington, 'Balloon "Taxi Service" to Take Satellites to Space' *BBC News* (25 September 2019) <https://www.bbc.com/news/uk-wales-49827415> accessed 10 August 2021.

[196] 'UK Space Conference: Drones and Satellite Cash Boost for Wales' *BBC News* (24 September 2019) <https://www.bbc.com/news/uk-wales-49801654> accessed 10 August 2021.

[197] ibid.

5.2.3 Prestwick Spaceport

A horizontal spaceport has been proposed at Glasgow Prestwick Airport in Ayrshire, Scotland.[198] In November 2020, both the UK and Scottish governments made a commitment to invest over £80.000.000 to develop horizontal launch capabilities there,[199] taking advantage of its 2.986 m runway[200] and the local aerospace business infrastructure.[201] The spaceport operators are in 'advanced discussions with several launch service providers', and hope to facilitate suborbital launches in 2022, followed by orbital launches in 2023.[202] The site is also a possible location for future space tourism.[203]

5.3 Vertical and Horizontal

5.3.1 Spaceport Machrihanish

Located at Campbeltown Airport in Argyll, Scotland, the Machrihanish site has much of the infrastructure needed for horizontal launches in place, including specialist fuelling facilities, transportable launch support and range safety systems,[204] and a 3,049 m runway—which was previously a back-up emergency landing site for the Space Shuttle.[205] Space technology companies QinetiQ and Telespazio VEGA UK signed MoUs with the spaceports operators in 2017,[206] but little progress has been made since then.[207] Nevertheless, the site aims to facilitate small suborbital rocket launches from 2021, and orbital launches from 2027.[208]

In sum (and as summarised in Table 1): it is clear that each of the seven planned UK spaceports are in different stages of development. SaxaVord, Space Hub Sutherland and Spaceport Cornwall would seem to offer the most promise for launches in the near-future, though it should be noted that legal obstacles remain, in the form of public and private opposition on environmental grounds.

---

[198] LaunchUK (n 10) 16–7.

[199] Williamson (n 155) 32.

[200] BBC News (n 126).

[201] LaunchUK (n 10) 16–7.

[202] ibid.

[203] Read (n 13) 17.

[204] LaunchUK (n 10) 15.

[205] Read (n 13) 18.

[206] 'New Milestone Reached in UK Spaceport Development' (*Discover Space UK*, 26 January 2017) <http://www.discoverspaceuk.com/news-details.php?id=938> accessed 10 August 2021.

[207] Williamson (n 155).

[208] LaunchUK (n 10) 15.

Table 1 Proposed UK spaceports (adapted from LaunchUK's 2020 report [n 9] with additional information added regarding on-going legal challenges)

	Launch mode	Payload capacity	Planned first launch	Customers	Legal obstacles
SaxaVord Spaceport	Vertical	≤1000 kg	2022	Lockheed Martin; ABL Systems	Consent withheld by a statutory agency on the grounds of historical importance—on-going; licence pending
Sutherland Space Hub	Vertical	<500 kg	2022	Orbex	Environmental opposition; planning permission currently being challenged in court; licence pending
Spaceport 1	Vertical	Suborbital <100 kg; orbital <350 kg	Sub-orbital: 2021; orbital: 2023	To be determined	Environmental opposition; licence pending
Spaceport Cornwall	Horizontal	≤300 kg	2022	Virgin Orbit	Licence pending
Spaceport Snowdonia	Horizontal	≤200 kg	2022	Multiple research projects	Licence pending
Prestwick Spaceport	Horizontal	To be determined	Sub-orbital: 2022; orbital: 2023	To be determined	Licence pending
Spaceport Machrihanish	Horizontal and vertical	To be determined	Sub-orbital: 2021; orbital: 2027	To be determined	Licence pending

6 Spaceport Collaboration Between the UK and the Rest of Europe

6.1 Political Concerns

In his discussion on political considerations surrounding spaceport placement, Thomas G. Roberts of the CSIS Aerospace Security Project cites three primary factors: accessibility, stability and public awareness.[209] In the following section, these factors will be briefly examined in relation to UK spaceports from the European perspective, before discussing on-going and future areas of collaboration:

[209] Roberts (n 3) 7.

1. Accessibility. As discussed, elements of Copernicus which required SSO and polar orbits were launched from Plesetsk Cosmodrome, over 2.500 kms from Brussels, and relatively difficult to transport expensive hardware to. By comparison, both the Sutherland and SaxaVord spaceports will be less than half the distance, at approximately 1.000 and 1.180 kms respectively. Moreover, the logistical infrastructure already in place between the UK and the rest of Europe means that the transportation of critical instruments from manufacturing centres will be comparatively easy and more economical.
2. Stability. Though the level of political instability caused by Brexit and the withdrawal negotiations was not insignificant by any means, it seems unlikely that political instability will threaten UK spaceports or the ability of domestic and European actors to make use of them. However, the public opposition to the Scottish vertical spaceports should be accounted for when considering this. In 2017, thousands of protesters blocked the entry to the spaceport in French Guiana, almost precluding ESA's access to the site.[210] The protest (which lasted more than three weeks) was to bring attention to local issues in the country.[211] Though not at the same scale, the environmental activist group Extinction Rebellion[212] staged a series of public protests in Scotland regarding the planned spaceports there, and the prospect for future protests and public opposition should be noted.
3. Public awareness. Whereas during the Cold War, spaceports were often state secrets,[213] in the modern world, they are frequently a site of tourism, and their visibility can be used to promote space programs and initiatives, and to boost the local economy. Indeed, making use of spaceports in the UK may have a dual function for Europe: not only would facilities nearby provide a tourism opportunity for both British and European citizens, but it may also help to foster a greater sense of cooperation between the two political entities. Thus, European use of UK spaceports may take on a normative, as well as an imperial, dimension.

6.2 On-Going Collaborations

An encouraging sign for future collaboration on UK spaceports concerns ESA's Boost! program, which provides 'co-funding, tailored expert advice and the use of testing facilities to help entrepreneurs to develop and prepare new space transportation services.'[214] The UKSA invested £12.000.000 into Boost! in 2019, which

[210] ibid.

[211] ibid.

[212] Ross J, 'Extinction Rebellion Joins Fight against Highland Space Port Plan' (*Press and Journal*, 30 June 2020) <https://www.pressandjournal.co.uk/fp/news/highlands/2297765/extinction-rebellion-joins-fight-against-highland-space-port-plan/> accessed 10 August 2021.

[213] Roberts (n 3).

[214] 'Boost! – ESA's Commercial Space Transportation Services and Support Programme' (*ESA*, May 2021) <https://www.esa.int/Enabling_Support/Space_Transportation/Boost!_ESA_s_Commercial_Space_Transportation_Services_and_Support_Programme> accessed 8 August 2021.

was one of the largest investments from any ESA member state.[215] As part of this initiative, ESA signed two new contracts in March of 2021 with UK-based rocket companies, Orbex and Skyrora, in order to 'help establish the UK as a new space transportation hub, offering launch services to low Earth orbit.'[216] The contracts total €7.000.000 of co-funding to Orbex for UK-based activities, and €3.000.000 to Skyrora.[217] As discussed in Sect. 5.1.2, Orbex plans to offer small satellite launches to LEO from Space Hub Sutherland from 2022,[218] and it was also the first UK-based space company to receive funding from the European Commission's Horizon 2020 fund, through which it was awarded €2.500.000.[219,220] Skyrora plans to offer launches to low Earth, sun-synchronous and polar orbits for payloads of up to 315 kg, and aims to begin commercial launches from a UK-based spaceport in 2022.[221]

In July of 2021, ESA signed another contract totalling £1.600.000 with the UK-branch of D-Orbit,[222] an Italian company which offers space logistics and transportation services.[223] The company plans to establish a 'satellite assembly, integration, and testing facility' at the Cornwall spaceport, in cooperation with a wide range of UK-based launch providers (including Virgin Orbit and Skyrora)[224] as well as with other UK and EU spaceports.[225] Ian Annett, Deputy CEO of the UKSA, said of the new contract:

> This funding is excellent news for D-Orbit and for the UK space sector. Their standardised end-to-end transportation services will promote increased cohesion between European spaceports and launch providers, encouraging greater international collaboration. This support,

---

[215] 'New Launch + On-Orbit Services To Issue From D-Orbit Under ESA's BOOST! Project—SatNews' (*SatNews*, 21 July 2021) <https://news.satnews.com/2021/07/21/new-launch-on-orbit-services-to-issue-from-d-orbit-under-esas-boost-project/> accessed 8 August 2021.

[216] 'ESA Boost for UK Space Transportation Initiatives' (*ESA*, 24 March 2021) <https://www.esa.int/Enabling_Support/Space_Transportation/ESA_boost_for_UK_space_transportation_initiatives> accessed 8 August 2021.

[217] ibid.

[218] ibid.

[219] Jeff Foust, 'Isar Aerospace, Orbex Raise Funding for Small Launch Vehicles' (*SpaceNews*, 10 December 2020) <https://spacenews.com/isar-aerospace-orbex-raise-funding-for-small-launch-vehicles/> accessed 8 August 2021.

[220] 'Orbex' (*Orbex*, 10 December 2020) < https://orbex.space/news > accessed 8 August 2021.

[221] 'Rocket Startup Skyrora Achieves a Successful Sub-Orbital Launch from Scottish Island' (*TechCrunch*, 15 June 2020) <https://social.techcrunch.com/2020/06/15/rocket-startup-skyrora-achieves-a-successful-sub-orbital-launch-from-scottish-island/> accessed 8 August 2021.

[222] 'ESA's Boost! Fosters New Launch and in-Orbit Services' (*ESA*, 20 July 2021) <https://www.esa.int/Enabling_Support/Space_Transportation/ESA_s_Boost!_fosters_new_launch_and_in-orbit_services> accessed 8 August 2021.

[223] SatNews (n 222).

[224] ESA (n 223).

[225] SatNews (n 222).

> alongside our world-class talent in the space sector, means the UK will be well-placed to exploit the opportunities that commercial launch from the UK will bring.[226]

Thilo Kranz, ESA's Commercial Space Transportation Program Manager, commented that the new contract would stimulate the European economy and help support space commercialisation, in line with ESA's Agenda 2025.[227] D-Orbit aims to promote logistical coordination and process standardisation between European spaceports,[228] and the inclusion of UK spaceports in this regard is significant.

Thus, although cooperation between the UK and Europe in other fields of the space sector may be complicated by Brexit, spaceport usage appears to be a strong, fruitful candidate for future opportunities, as evinced by on-going ESA support for relevant initiatives, and the promotion of coordination and standardisation by companies such as D-Orbit. The European space sector may therefore be in an ideal position to take full advantage of the opportunities which upcoming UK spaceports will offer, in a manner that will be mutually beneficial to all parties.

7 Conclusion

As has been discussed, the opening of several new vertical and horizontal spaceports in the UK represents an invaluable opportunity—not just for their domestic space sector, but for the rest of Europe as well. The proximity to the rest of the continent, combined with its position as a northern island means it is ideal for small satellite launches into polar and sun-synchronous orbits, which are only set to become more valuable—with one recent report indicating that small satellite payloads being launched into LEO will increase tenfold by 2028.[229] That said, the spaceports currently being proposed are all at different stages of development, and there are legal obstacles that still need to be overcome before they can open. Nevertheless, satellite and launch companies based in Europe are already making use of the potential for spaceports in the UK, and several partnerships have been formed in anticipation of launches in the near future. In addition, launches for ESA missions on UK soil, despite the UK's exclusion from certain EU-funded programs, seem to be a realistic prospect in the short to medium term, although this will be affected by

---

[226] D-Orbit, 'Press Release: D-Orbit signs contract with the European Space Agency' (2021) <https://skyandspace.co/wp-content/uploads/2021/07/D-Orbit-Contract-with-European-Space-Agency.pdf> 1.

[227] ibid 2.

[228] ibid 1.

[229] 'Small Satellite Market Size, Growth & Global Forecast [2021–2028]' (Fortune Business Insights 2021) Market Research Report <https://www.fortunebusinessinsights.com/industry-reports/small-satellite-market-101917> accessed 12 August 2021.

the opening of additional spaceports on the continent. Furthermore, the continuation of this relationship may ultimately be a function of the UK's continued membership of ESA, which itself may be affected by further integration with the EU.

Connor Hogan holds a bachelor's degree in politics, Philosophy and Economics (PPE) from Queen's University, Belfast (QUB), for which he completed a dissertation on public and private initiatives in the space industry. He is currently studying for an MSc in Politics at the School of Politics and International Relations (SPIRe) in University College Dublin (UCD) and is undertaking a thesis on labour regulations in a future space-based economy.

UK Spaceports and Launch Services: An Overview of the Assessment of Environmental Effects and Environmental Impact Assessment

Nancy Riordan

Abstract The United Kingdom (UK) is positioning itself to become the first country to launch into orbit from Europe by 2022. The UK's 2018 Space Industry Act established the legal framework for commercial launch activity with regulatory functions delegated to the Civil Aviation Authority (CAA). On 24th May 2021, the UK's space industry regulations were presented to parliament specifying the rules and regulations which must be followed by spaceport and launch operators. They came into effect on 29th July 2021. The rules and regulations include licensing requirements, environmental considerations, liability and insurance, security, and an accident investigation process. This chapter will explore the environmental considerations required for licensing UK spaceports and launch activities by asking: what is the role of assessment of environmental effects (AEE) and environmental impact assessment (EIA) in the regulation of proposed UK spaceports and launch services and what are potential transboundary (TEIA) challenges? This chapter begins with a brief background on UK spaceports and the environmental aspects of the legal and regulatory process for proposed spaceports and launch services. Part two will discuss the details of EIA and TEIA and how they apply. Part three will present data on the status of EIAs for UK spaceports. Concluding remarks will summarize findings, propose future research, as well as draw attention to applicable UN COPUOS Long-term Sustainability (LTS) guidelines and the need for more attention to EIA and TEIA as tools for realizing aspects of the long-term sustainability of space activities. The research for this chapter was conducted through desk research utilizing publicly available documents and other literature. The purpose of this research is to explain the multi-level environmental governance challenges for UK spaceports and launch services and how EIA applies at all levels, from global to local.

N. Riordan (✉)
Global Governance and Human Security, McCormack Graduate School of Policy & Global Studies, University of Massachusetts, Boston, USA
e-mail: nancy.riordan@umb.edu

A. Froehlich (ed.), *Spaceports in Europe*, Studies in Space Policy 34,
https://doi.org/10.1007/978-3-030-88311-9_5

1 Background

This chapter begins with a brief background on UK spaceports and the environmental aspects of the legal and regulatory process for proposed spaceports and launch services. Under the UK Space Industry Act of 2018, a spaceport is defined as "any site from which a spacecraft or carrier aircraft is intended to launch is considered a spaceport and must be licensed. A site at which controlled and planned landings of spacecraft are to take place is also a spaceport and must be licensed."[1] Similarly, in the U.S. Federal Aviation Authority (FAA) Reauthorization Act of 2018, the term spaceport means "a launch or reentry site that is operated by an entity licensed by the Secretary of Transportation."[2] The U.S. FAA-AST Office of Commercial Space Transportation has played a significant role in assisting the UK prepare for launch and U.S. company Lockheed Martin is set to be one of the first commercial companies to launch from Scotland's Saxa Vord Spaceport (renamed from Shetland Space Center in June 2021).[3] Hand in hand with the U.S., UK draft legislation has stated "for UK launch to take place, government intervention is required to put in place high level powers to develop a comprehensive regulatory framework to enable potential US-based operators the regulatory certainty they need to plan for expansion in the UK and to ensure that the experimental nature of spaceflight is adequately reflected in UK legislation."[4] According to the European Space Policy Institute's *ESPI Yearbook 2020*, many spaceport projects are being planned by European countries and agreements have been signed for the construction or operation of new spaceports.[5] Several motives are driving countries to consider the construction of a spaceport. In the UK, spaceports are being promoted as initiatives that can support economic growth, job

---

[1] Space Industry Act 2018 Chapter 3(2)(a)(b).

[2] Pub. L. 115–254, div. B, title V, §580(b)(1), Oct. 5, 2018, 132 Stat. 3395.

[3] Memorandum of Cooperation—FAA and United Kingdom, signed July 2014 https://www.faa.gov/space/additional_information/international_affairs/media/Memorandum_of_Cooperation_FAA_and_UK_AST_DFT-UKSA-CAA_signed_July-2014.pdf, Shetland Space Center "Name change and rebrand for our rocket site," 30 June 2021 https://shetlandspacecentre.com/general/name-change-and-rebrand-for-our-rocket-site/, "British government works closely with FAA to finalize commercial launch regulations," *Geospatial World*, 27 July 2020, https://www.geospatialworld.net/news/british-government-works-closely-with-faa-to-finalize-commercial-launch-regulations/, Alan Boyle, "Lockheed Martin wins $31 M from UK Space Agency for spaceport in Scotland," *Geek Wire*, 16 July 2018 https://www.geekwire.com/2018/lockheed-martin-31m-scotland-spaceport/.

[4] Drat UK Space Industry Bill—Spaceflight, Reference No: RPC-3515(1)-DfT, p. 5, The Space Industry Bill received Royal Assent and became the Space Industry Act on 15 March 2018 https://www.legislation.gov.uk/ukpga/2018/5/pdfs/ukpgaod_20180005_en_001.pdf, See also US-UK Spaceports Agreement 16 June 2020, the Agreement in the form of an Exchange of Notes between the Government of the United Kingdom of Great Britain and Northern Ireland and the Government of the United States of America on Technology Safeguards associated with United States Participation in Space Launches from the United Kingdom, https://assets.publishing.service.gov.uk/government/uploads/system/uploads/attachment_data/file/927083/CS_UK_USA_1.2020_USA_Techno_Safeguards_Space_Launches.pdf.

[5] ESPI Yearbook 2020 Space policies, issues and trends, June 2021, 65.

creation, skills development, and attract investments.[6] Over the last decade, the UK has openly stated its ambition to become a launching state, with the long-term goal of being able to support suborbital operations and orbital delivery of small satellites.[7] In August 2012, the Department for Transport and UK Space Agency tasked the Civil Aviation Authority (CAA) to conduct a detailed review of operational and regulatory requirements to enable spaceplanes to operate from the UK by 2018.[8] Other government led studies have investigated the feasibility of locations for UK spaceports and vertical and horizontal launch, for northern Scotland in particular. These include the 2014 Department for Transport 'Consultation on criteria to determine the location of a UK spaceport,' which surveyed opinions on the strategic position to be adopted by government on the location of a spaceport and the potential for existing airports and ranges to be converted to a spaceport for horizontal launch.[9] That consultation as well as the July 2014 CAA led 'Review of commercial spaceplane certification and operations' technical report used the framework of existing environmental policy and legislation that applies to aviation, and considered its applicability to spaceplanes and spaceports. The 2014 CAA report also acknowledged that since "spaceplanes would be considered aircraft, for at least part of the journey, aviation environmental regulations would apply to spaceplane operations."[10] Regarding a spaceport that can support vertical launch operations that reach orbit, the 2017 SCEPTRE study conducted by Demios Space UK Ltd. found that the UK is geographically suited for launch to Polar and Sun—Synchronous Orbits. More specifically, for reasons of flight safety and human casualty risks, the study determined the north of Scotland as the only feasible region for vertical launch using one particular model of operation: vertical launch of a Firefly Alpha as a 'reference vehicle' an imported U.S. launch

[6] Ibid.

[7] The U.K. is poised to become the first country in the region to achieve orbital launches. According to the Guidance Document for Spaceport License Applicants and Spaceport Licensees, the regulator (CAA) will use the International Standard Atmosphere (47 km) as the stratopause (i.e., the upper limit of the stratosphere) for the purposes of determining whether an activity is 'sub-orbital' or orbital, the U.S. FAA rules that "orbital spaceflight occurs when a spacecraft is placed on a trajectory with sufficient velocity to place it into orbit around the Earth. Suborbital spaceflight occurs when a spacecraft reaches space, but its velocity is such that it cannot achieve orbit. Many people believe that to achieve spaceflight, a spacecraft must reach an altitude higher than 100 km (62 miles) above sea level https://www.faa.gov/space/additional_information/faq/#cl6, in Europe sub-orbital launches have taken place at Andøya island spaceport in Norway, Esrange spaceport in Sweden, and most recently from Shetland space center by private space company Skyrora and their Skylark Nano rocket that reached an altitude of six kilometers.

[8] UK Government Review of Commercial Spaceplane Certification and Operations technical report July 2014.

[9] Supporting Commercial Spaceplane Operations in the UK Consultation on Criteria to Determine the Location of a UK spaceport 2014 study location p. 5., At the time, no consulting occurred to gather local communities' and stakeholders' preferences. The consultation was based on pre-existing environmental regulations for international aviation covering issues such as noise, air quality (including carbon emissions) and the storage of hazardous materials.

[10] Ibid, 12.

vehicle capable of carrying payloads of up to 500 kg.[11] The 2017 SCEPTRE study recommended three sites: 1st choice considered the Shetland Isles to have the best orbital access, but logistically the most challenging, 2nd the Mhoine peninsula "a large uninhabited area" logistically feasible but populations downrange in the Faroes and Iceland "constrain the available trajectories and reduce the mass that can be delivered to orbit," and 3rd Scolpaig on North Uist has previous defense department range infrastructure, but not able to meet the U.S. FAA – AST casualty rate requirements.[12] The report also highlights that while the UK Parliament leads on the legislation to enable space activities in the UK, there are significant powers that are devolved to the Scottish Parliament for instance. These include matters of the environment and land use planning (led in Scotland by the relevant local planning authorities (LPA) and discussed in the next section). The study suggests that "while UK legislation will be taken forward through the Modern Transport Bill it would not be expected that the UK Parliament would legislate on these matters that affect Scotland without the consent of the Scottish Parliament."[13] The study also noted that both Scottish Natural Heritage and Scottish Environmental Protection Agency would need to be consulted prior to any application being considered. Most recently, an 'Explanatory Memorandum to the Space Industry Regulations 2021' makes clear that "relying on sites in other countries to launch UK satellites and scientific experiments into space, where timely launch slots are decreasing and launch costs are increasing" are driving the UK ambitions to build spaceports.[14] Moreover, "[t]here are also often extensive legal, export and regulatory processes required to launch from foreign sites. This means access to space is becoming a barrier to growth for the UK's space industry, particularly for the UK's small satellite manufacturers and businesses in the UK space sector that rely on these small satellites to provide their services and products."[15] Until now, the UK government and commercial space industries have relied on spaceports in other countries to launch satellites into space. Most recently, the UK firm OneWeb launched thirty-four satellites from Baikonur, Kazakhstan on a Russian Soyuz rocket in 2021 and six satellites from the Guiana Space Centre in Kourou, French Guiana, in 2019.[16]

[11] SCEPTRE – DMU – FR Final Report 17 February 2017 https://www.hie.co.uk/media/6626/sceptre-final-report-february-2017.pdf.

[12] Ibid, 21.

[13] Even though the process began in 2012 and the SCEPTRE study was published in 2017 stakeholders and the public had yet to be consulted. SCEPTRE advised that prior planning about how any licensing arrangement with UK Space Agency would engage with these relevant bodies was needed. Also, the "acceptance of any local community cannot be assumed, and considerable effort and care should be taken to make sure that a transparent and honest discussion is undertaken with both the local community and particularly any crofting interest that may have an impact on the project.".

[14] Explanatory Memorandum to The Space Industry Regulations 2021, 4, See also, 2016 Space Industry Bill Spaceflight Impact Assessment, 4, 5.

[15] Ibid.

[16] "OneWeb launches 34 satellites from Kazakh cosmodrome in global internet push," *Reuters*, 7th February 2020, https://www.reuters.com/article/us-space-exploration-oneweb-launch/oneweb-launches-34-satellites-from-kazakh-cosmodrome-in-global-internet-push-idUSKBN20113P.

Only one British launch vehicle has ever been used for orbital space flight—Black Arrow, and that was tested at the Australian Woomera Test Range in 1971.[17] A full history of UK space activity is beyond the scope of this chapter. However, a European Space Agency paper prepared by Douglas Millard (2005) "An overview of United Kingdom space activity 1957–1987" provides background and a comprehensive bibliography of notable historical research.[18]

The map in (Fig. 1) shows the seven locations that have been identified for commercial spaceports with five in Scotland (Hebrides, Shetland, Sutherland, Ayrshire and Argyll), one in Gwynedd, North Wales and one in Cornwall. Three in the north of Scotland can host vertical launches and four others will be designated for horizontal launches.[19] Several of these spaceports are planning to be operational by 2022–2023.

1.1 The Environmental Aspects of the Legal and Regulatory Process for Proposed Spaceports and Launch Services

The role of assessment of environmental effects (AEE) in the regulation of proposed UK spaceports and launch services can be found in section eleven of the UK Space Industry Act 2018 "Grant of licenses: assessment of environmental effects," and applies to both applications for a spaceport license and an operator license authorizing launches of spacecraft or carrier aircraft.[20] The launch license applicant might involve domestic or international commercial entities and is separate from the spaceport license.[21] In the case of U.S. launch companies for example, they require additional

[17] Thomas G Roberts, *Spaceports of the World* (CSIS Report, 2019) https://aerospace.csis.org/wp-content/uploads/2019/03/190313__SpaceportsOfTheWorld.pdf.

[18] Millard (2005) suggests for a survey of the UK's space activities up to the early 1990s, see Jane's Space Directory 1993–94 (Wilson, 1993), parliamentary reports of 1967 (House of Commons, 1967), 1971 (House of Commons 1971) and 1987 (House of Lords, 1987), and Adams (1976), Massey and Robins (1986) for a detailed, first-hand account of the UK's formative and early space science program at home, with the United States and within Europe, also Krige and Russo (2000) and Krige, Russo and Sebesta (2000), Marsh, P. (1991), 'Britain in Space', in R. Nicholson, C. M. Cunningham and P. Gummett (Eds.), and Millard, D. (2001), The Black Arrow Rocket: A history of a satellite launch vehicle and its engines, (Science Museum, London 2001), Morton, Peter (1989), Fire Across the Desert: Woomera and the Anglo-Australian Joint Project, 1946–1980.

[19] As noted in the UK space center FAQ, "a vertical launch is …an upright rocket that ignites and lifts off straight up. Typically rockets will have 2–3 stages that ignite and then fall away when their fuel is used up. A horizontal launch means that a plane takes off from a runway, carrying a rocket with it. Once the plane is in the air and safely away from urban areas, the rocket can detach, ignite, and head into space. Horizontal launches save on fuel costs as the plane does most of the work getting the rocket into the upper atmosphere, but they're limited by how much the plane can carry so these types of launches tend to only be feasible for very small satellites" https://spacecentre.co.uk/blog-post/uk-spaceport-faq/.

[20] Space Industry Act 2018, Sect. 11, https://www.legislation.gov.uk/ukpga/2018/5/section/11/enacted.

[21] Ibid.

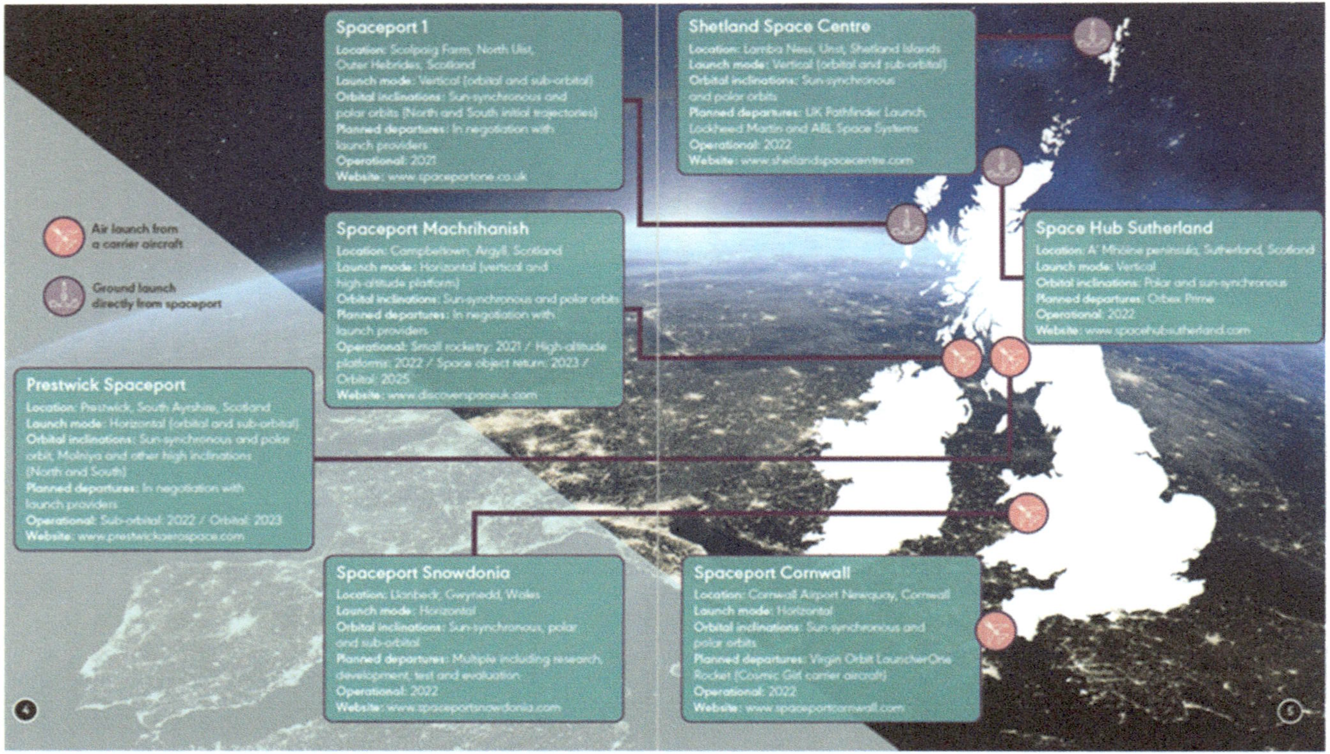

Fig. 1 Seven UK spaceports. *Source* UK Space Agency blog https://space.blog.gov.uk/2021/07/30/new-regulations-mean-lift-off-for-uk-spaceflight/

FAA licensing and a prior TSA was signed between the UK and U.S. governments to allow U.S. space companies to operate in the UK.[22]

The 2018 Space Industry Act provides specific guidance from the Secretary of State that the CAA *must not* [emphasis added] grant a license unless the applicant has submitted an assessment of environmental effects. An "assessment of environmental effects" in relation to a spaceport license, means "an assessment of the effects that launches of spacecraft or carrier aircraft from the spaceport in question, or launches of spacecraft from carrier aircraft launched from the spaceport, are expected to have on the environment; in relation to an operator license authorizing launches of spacecraft or carrier aircraft, means an assessment of the effects that those launches are expected to have on the environment."[23]

Space activities present a two-fold challenge with regard to environmental impacts. This duality involves our reliance on a multitude of space-based services especially for earth observation and communication to better understand our planet and climate change, yet commercial spaceflight in the UK will have environmental implications at global, national, regional, and local levels. It is acknowledged by the UK government that, "[s]paceflight has the potential to affect climate change, local air quality and noise levels which can impact human health, ecological receptors and the marine environment. To deliver the economic, social, and environmental objectives, a balance will need to be struck between avoiding or minimizing negative environmental impacts and maximizing the contribution of spaceflight activities to the economy and the environment."[24] The main environmental effects of UK spaceflight activities are likely to be: "the effects of spaceflight emissions on climate change and on levels of ozone in the upper troposphere and stratosphere, the effects on local air quality around the spaceport, the effects of spaceflight noise on people and wildlife at the spaceports and under flightpaths, and the effects on the marine environment from jettisoned objects, as well as coastal spaceports and launch activities."[25] The requirements for submitting an AEE are found in the CAA's 'Guidance for the Assessment of Environmental Effects': "[t]he purpose is to ensure applicants for a spaceport or launch licence have considered the environmental effects of their activities and if necessary taken (or identified) porportionate steps to avoid, mitigate

---

[22] US-UK Spaceports Agreement 16 June 2020, International treaty UK/USA: Agreement in the form of an Exchange of Notes between the United Kingdom and the United States of America on Technology Safeguards associated with United States Participation in Space Launches from the United Kingdom [CS USA No.1/2020].

[23] Space Industry Act 2018.

[24] Guidance to the Regulator on Environmental Objectives Relating to the Exercise of its Functions Under the Space Industry Act 2018 Given by the Secretary of State under Sect. 2(2)(e) of the Space Industry Act 2018, 8.

[25] Guidance for the Assessment of Environmental Effects, 2021, 9, see also, Jemma-Anne Lonsdale and Claire Phillips, *Space Launches, and the UK Marine Environment* (Marine policy 2021), J.A Dallas et al., *The Environmental Impact of Emissions from Space Launches: A Comprehensive Review* (Journal of cleaner production 2020), Martin Ross, et al., *Limits on the Space Launch Market Related to Stratospheric Ozone Depletion* (Astropolitics 2009), and Martin Ross, et al., *Potential climate impacts of black carbon emitted by rockets* (Geophysical Research Letters 2010).

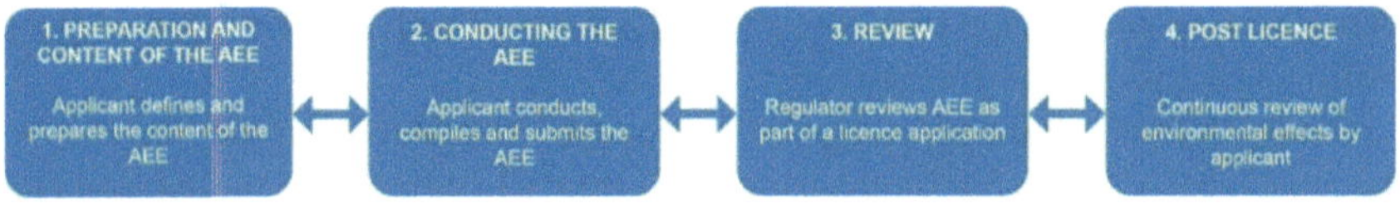

Fig. 2 AEE process. *Source* Guidance for the assessment of environmental Effects

or offset the risks and their potential effects."[26] The AEE is a process of identifying, quantifying, and evaluating the potential effects of the proposed space activities on the environment. The process occurs in four stages as shown in (Fig. 2).

The AEE must also address these specific environmental topics.[27]

- population and human health.
- biodiversity (for example, ecology, flora, and fauna).
- air quality.
- noise and vibration.
- water (for example, quality, and quantity).
- marine environment.
- climate (for example, greenhouse gas emissions, impacts relevant to adaptation).
- land, soils, and peat.
- landscape and visual impact.
- cultural heritage (including architectural and archaeological aspects).

'Guidance for the assessment of environmental effects' also requires that the AEE include a plan for assessing mitigation measures. The AEE must include a description of the mitigation measures, their expected effectiveness, and provide an assessment of any residual effects. Also, control measures must be designed to avoid, reduce, or compensate for any significant adverse environmental effects. Spaceport and launch license applicants are required to: "identify a range of control measures, in line with the mitigation hierarchy, explain the likely effectiveness of each measure, assess and provide a description of the significance of any residual effect, state as a result of the analysis, which measure(s) they intend to use, identify where monitoring is required to ensure the success/effectiveness of the measures, and provide sufficient information to demonstrate how the chosen measure(s) can be effectively implemented."[28]

---

[26] Guidance for the Assessment of Environmental Effects, 29 July 2021,14.

[27] Ibid, 22.

[28] Ibid, 26.

1.2 Global and Local Environmental Governance Considerations

1.2.1 Reserved and Devolved Matters

Following the UK's departure from the European Union (EU), the CAA as regulator and space industries themselves have needed to keep informed of all new environmental legislation, policies, and guidance. Since Brexit, "the UK government has generally kept the EU environmental rules adjusting them, when necessary, to fit the new legal reality."[29] Regulation of space remains a 'reserved' matter for the UK parliament but ground-based facilities and infrastructure for spaceports as well as launches involve matters that are applicable to devolved regulations and responsibility rests with the Scottish Government, Welsh Government and Northern Ireland Executive for environmental regulations and planning.

1.2.2 International (and National) Environmental Treaties, Guidelines, and Standards

The UK is a contributor, member, or signatory to the following international treaties and guidelines:

Environmental impact assessment is recognized in several international conventions and protocols.

- ESPOO—the Convention on Transboundary Environmental Impact Assessment and protocol on strategic environmental assessment (SEA)
- RAMSAR—the Convention on Wetlands of International Importance
- AARHUS—the Convention on Access to Information, Public Participation in Decision-making, and Access to Justice in Environmental Matters
- UNFCC—the United Nations Framework Convention on Climate Change
- UNCLOS—the United Nations Convention on the Law of the Sea
- The Protocol on Environmental Protection to the Antarctic Treaty

The marine environment: the Secretary of State has set an environmental objective for the regulator to 'protect the marine environment from the impacts of spaceflight activities' impacts on the marine environment extend beyond UK waters.[30]

International

- London Protocol
- Convention for the Protection of the Marine Environment of the North-East Atlantic (the 'OSPAR Convention')

---

[29] Jacquelyn Genevra and Emma Bichet, *Environmental Law After Brexit* (Environmental Journal 17th May 2021) https://environmentjournal.online/articles/environmental-law-after-brexit/.

[30] UK DOT Guidance to the Regulator on Environmental Objectives Relating to the Exercise of its Functions Under the Space Industry Act 2018, 29 July 2021, 22.

- United Nations Convention on the Law of the Sea (UNCLOS)

National

- Marine and Coastal Access Act 2009
- UK Marine Strategy Regulations 2010
- Marine Policy Statement (MPS)
- 25 Year Environment Plan and Maritime 2050

Climate: The Secretary of State has set an environmental objective for the regulator to 'minimize emissions contributing to climate change resulting from spaceflight activities'.[31]

International

- Paris Agreement

National

- Climate Change Act 2008
- Clean Air Strategy 2019
- Clean Growth Strategy
- 25 Year Environment Plan

Air: the adverse impacts of poor local air quality on ecological receptors due to emissions from spaceflight activities should be identified and conservation objectives considered.[32]

Internationally designated sites

- special protection areas (SPAs),
- special areas of conservation (SACs)
- Ramsar sites

Nationally designated sites

- sites of special scientific interest (SSSIs)
- areas of special scientific interest (ASSIs)
- marine conservation zones (MCZs)
- Marine protected areas (MPAs)
- national nature reserves, local nature sites such as local wildlife sites, ancient woodlands and national and local nature reserves and protected or priority habitats and species

Noise standards

International

- International Organization for Standardization (ISO)

---

[31] Ibid, 12.

[32] Ibid, 14.

- World Health Organization (WHO) National
- BSI British Standards

Space Debris

- Space Debris Mitigation Guidelines (COPUOS)
- Guidelines for the Long-term Sustainability of Outer Space Activities (COPUOS)
- European Code of Conduct for Space Debris Mitigation
- Inter-Agency Debris Coordination Committee Space Debris Mitigation Guidelines
- Space Debris Mitigation Requirements BS ISO 24113–2019
- Disposal of orbital launch stages BS ISO 16699:2015
- Requirements on Space Debris Mitigation for European Space Agency Projects

While the debris guidelines are not legally binding on the UK or any other signatory, the Space Industry Act 2018 places responsibility on the regulator (CAA) to take account international space debris mitigation guidelines when carrying out its functions under Sect. 2(2)(h), and the Space Industry Regulations 2021, give the regulator the power to place conditions on any license to address debris mitigation (Schedule 1, paragraph 1(g)) "These provisions have been included to ensure the safety, security and sustainability of the orbital environment. The regulator's duties with regard to debris mitigation sit alongside its duty to take account of the environmental objectives set by the Secretary of State."

2 EIA and TEIA and How They Apply to UK Spaceports and Launch Services

The previous section discussed the AEE for UK spaceport and launch licensing. AEE and environmental impact assessment (EIA) are terms often used interchangeably but are not the same thing. Think of the AEE as the final dossier of information for the CAA about environmental effects that accompanies the license application and the EIA as the methodology and methods used to build that assessment in report that is submitted to a local planning authority (LPA) first. The 2021 'Guidance for the assessment of environmental effects' document states that there are established frameworks for conducting environmental impact assessments. Specific to the UK, EIA is required by the 2017 Town and Country (EIA) Regulations and the 2017 Marine Works (EIA) regulations and the Chartered Institute of Ecology and Environmental Management (CIEEM) Ecological Impact Assessment Guidelines.[33]

[33] Guidance for the Assessment of Environmental Effects, 29 July 2021, 14. The guidance also states in Sect. 3 the AEE may be required to go beyond the scope of these existing frameworks to reflect the nature of space activities.

2.1 What is an Environmental Impact Assessment (EIA)?

According to a 2018 UN Environment report, environmental impact assessments (EIA) are the "most commonly known, used, and globally widespread, environmental planning and management tools."[34] The report emphasized that "the objective of these tools is to make sure that all critical information to predict future impact on the environment is supplied and considered in the decision-making process."[35] The goal of EIA is to avoid the implementation of any activity with significant negative impacts on the environment while enhancing the activity's positive impacts. EIA began as a broad process that emerged from the National Environmental Policy Act 1970 (NEPA) in the United States.[36] Over the past 25 years, EIA requirements have been extended to scores of countries, levels of government, and to private sector proposals.[37] In 1992, the Rio Declaration on Environment and Development determined, in Principle 17, that EIA should be the tool for achieving sustainable development. EIA have been defined as "the process of identifying, predicting, evaluating, and mitigating the biophysical, social, and other relevant effects of development proposals prior to major decisions being taken and commitments made."[38]

The EIA process involves several steps and the steps in the process can vary considerably. For example, the legislation within in some countries require post-decision monitoring while others do not. The order of the steps in the process may also vary.[39]

The basic procedural steps of EIA include:

- Screening: to determine which activities may have significant environmental impacts
- Scoping: to identify which potential impacts are relevant to assess and to find alternative options such as project locations, scales, operating conditions and a 'no action' option
- Assessing and evaluating impacts (adverse and beneficial) and the development of alternatives
- Mitigation involves the introduction of measures to avoid, reduce, remedy, or compensate for any significant adverse impact
- Public consultation and participation aim to ensure the quality, comprehensiveness and effectiveness of the EIA and the public's views are included in the decision-making process

---

[34] UNEP, *Assessing Environmental Impacts: A Global Review of Legislation* (Nairobi Kenya 2018).

[35] Ibid, 2.

[36] 1970 National Environmental Policy Act (NEPA).

[37] Ortolano and Shepherd, *Environmental Impact Assessment: Challenges and Opportunities* (Impact Assessment Vol 13 1995).

[38] Senécal, et al., *Principles of Environmental Impact Assessment* (Fargo North Dakota 1999).

[39] Glasson, et al., *Introduction to environmental impact assessment* (New York Routledge 2012).

- Reporting of the Environmental Impact Statement (EIS) is a vital step in the process
- Monitoring, compliance, enforcement, and environmental auditing

For EIA in the UK, a Friends of the Earth report documents how EIA is required for some types of development, but not others and deciding whether an EIA is required ("screening") can be a source of major dispute between developers, communities, and local authorities.[40] When undertaking screening, the regulations define two different types of projects, those where EIA must take place or where it may be needed.

For the UK specifically (see Fig. 3), these are set out in two Schedules as follows:

> Schedule 1 projects: an EIA must always be carried out (except where regulations 60–63 apply such as where the 'sole purpose' of the proposals is either military/national defense or in response to a civil emergency). Schedule 2 projects: an EIA must be carried out if the development is likely to have a significant impact on the environment by virtue of its nature, size, or location. Unfortunately, the definitions allow for considerable uncertainty about the need for EIA in specific circumstances and this can result in legal challenge.[41]

A developer for a spaceport may voluntarily submit an EIA to a local planning authority (LPA) with information whether they believe the project (depending on whether a new launch site is being built or the project is renovating an existing range previously belonging to the department of defense) under Schedule 1 or Schedule 2.[42] However, it is the local planning authority who makes decides if an EIA is required in consultation with the applicant. "The local planning authority (LPA) will likely be asked by the applicant whether EIA is necessary for a Schedule 2 submission. Regulations require the LPA to provide on request a formal opinion as to whether EIA should be carried out."[43]

2.2 EIA in a Transboundary Context (TEIA)

The previous section defined EIA and presented a brief overview of how the process is carried out in the UK. The EIA process is not a panacea for environmental challenges. However, there are important benefits, including the public nature of the process through consultation and documentation as well as facilitating an informed decision at the regulatory level. Such public involvement, "may serve as testimony to good governance in general, or, in more practical terms, be considered improved planning due to the concrete contribution of the public to the decision-making process...[and]

[40] Friends of the Earth, UK *Environmental Impact Assessment* September 2020.

[41] Ibid, 4.

[42] Ibid, 9.

[43] Ibid, see also the EIA process under Scottish law https://www.gov.scot/policies/environmental-assessment/environmental-impact-assessment-eia/ and Welsh law https://naturalresources.wales/guidance-and-advice/business-sectors/planning-and-development/advice-for-developers/environmental-assessment/?lang=en.

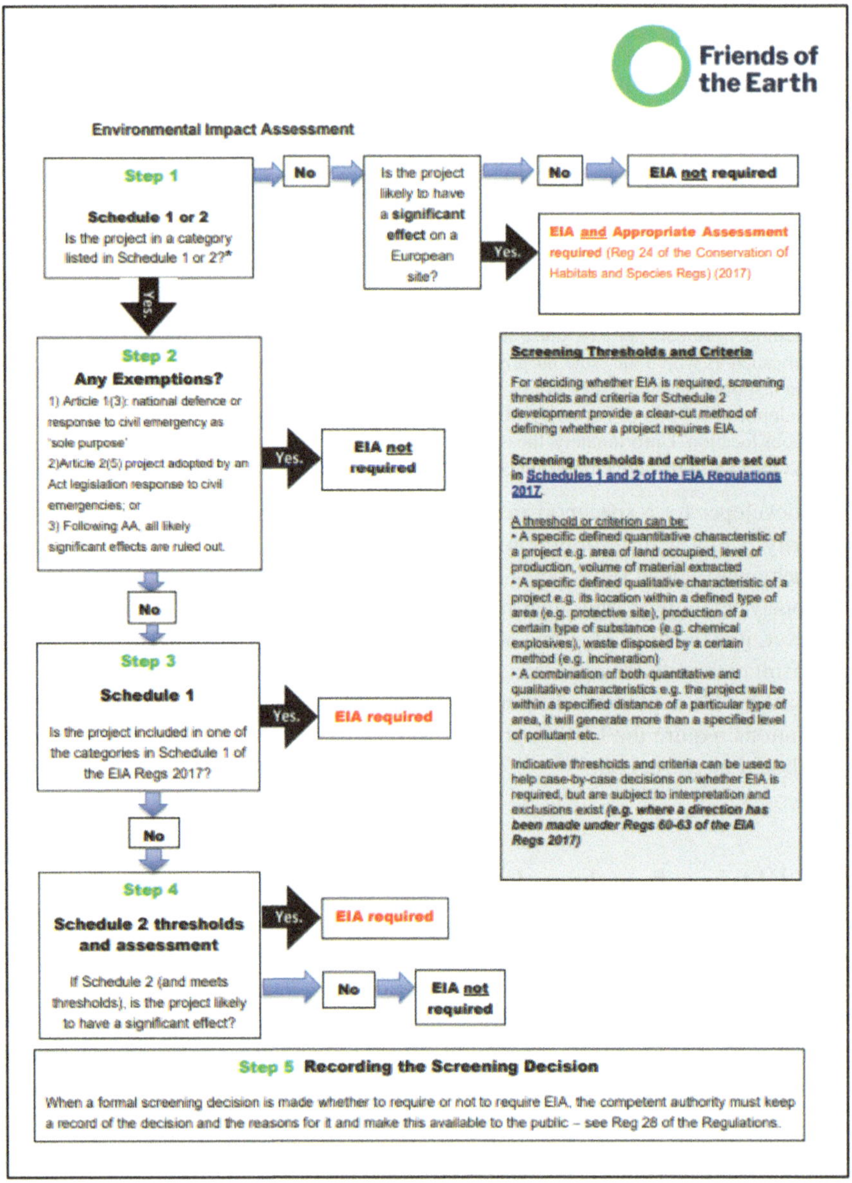

Fig. 3 UK EIA process. *Source* Friends of the Earth, EIA 2020

it is also a tool for reducing conflict and developing wider support for the eventual decisions."[44] EIA enables the public to gain information about planned activities and have collective influence.[45]

This next section will address EIA in a transboundary context. EIA in a transboundary context (TEIA) is defined as an EIA between territorial states.[46] The UK has ratified the 1991 UNECE Convention on Environmental Impact Assessment in a Transboundary Context, known as the Espoo Convention, that has specific rules for conducting an EIA of activities located on the territory of one contracting party, defined as the Party of origin, and likely to cause significant adverse transboundary impact in another contracting party, defined as the affected Party (Article 2).[47] TEIA can also be understood in broader terms as "encompassing EIA procedures designed to evaluate possible impacts by human activities on another state's environment or the environment of areas beyond a state's national jurisdiction" (ABNJ). There is specific guidance on TEIA in the UK government 'Guidance for the assessment of environmental effects' document.[48] The instructions are clear in Sect. 4.3 an AEE "must describe how specific potential effects have been identified and, where possible, quantified. When considering the significance of effects, it must cover the direct effects and any indirect, secondary, cumulative, transboundary, short term, medium term, long term, permanent and temporary, positive and negative effects of the proposed spaceflight activities." More specifically for launch operations, Sects. 4.57–4.61 environmental effects must be considered for the entire launch and the applicant must identify the vicinity of the launch, the trajectory—areas along the launch/return trajectory and drop zones such as the areas where any jettisoned objects will be deposited into the sea. The AEE must also provide an assessment of whether there could be any global or ABNJ impacts from the proposed activities, for example climate change impacts. "It is possible that proposed activities could affect environmental features in several areas of sovereignty and across national boundaries. The AEE must cover the environmental zone of influence, regardless of administrative boundaries, to take into account transboundary effects. If the applicant has insufficient information available to properly define the environmental zone of influence, this must be acknowledged in the AEE, and a precautionary approach must be taken."[49]

---

[44] Lotta Viikari, *Environmental Impact Assessment and Space Activities* (Advances in Space Research 2004).

[45] Richard K Morgan, *Environmental impact assessment: a methodological perspective* (Dordrecht: Kluwer Academic 1998), An abundance of important EIA literature can also be drawn on that debates the effectiveness of EIA (Morgan, 2012; Bond et al., 2012; and Cashmore et al., 2004, 2010).

[46] Bastmeijer and Koivurova, *Theory and Practice of Transboundary Environmental Impact Assessment* (Leiden, Martinus Nijhoff 2008) and Druel, *Environmental Impact Assessments in areas beyond national jurisdiction* (IDDRI 2013).

[47] Convention on Environmental Impact Assessment in a Transboundary Context (Espoo Convention) text of the Convention, 2017 (ECE/MP.EIA/21/Amend.1). https://unece.org/environment-pol icyenvironmental-assessment/text-convention.

[48] Guidance for the Assessment of Environmental Effects, 2021 p. 23.

[49] Ibid, 23.

In the ESPOO convention there are six key steps for the TEIA process[50]:

1. Notification and transmittal of information, Article 3.
2. Preparation of the EIA information/report by the spaceport developer, Article 4 and Appendix II.
3. Public participation, dissemination of information and consultation, Articles 3.8, 2.2, 2.6 and 4.2
4. Consultation between concerned Parties, Article 5.
5. Examination of the information gathered and final decision, Article 6.1
6. Dissemination of information on the final decision, Article 6.2

TEIA for launches from spaceports in Scotland will be discussed in the next section. Specifically for Scotland TEIA involves both the 1991 UNECE Convention on Environmental Impact Assessment in a Transboundary Context, Espoo Convention and the Town and Country Planning (Environmental Impact Assessment) (Scotland) Regulations 2017. These Regulations include a clause for consideration of transboundary impacts where a "development proposed to be carried out in Scotland is the subject of an EIA application and is likely to have significant effects on the environment in an EEA State other than the United Kingdom" then Scottish Ministers must send to the EEA state the particulars mentioned in paragraph (3) which are: "(a) a description of the development, together with any available information on its possible significant effect on the environment in another EEA State; and, (b) information on the nature of the decision which may be taken."[51]

3 Status of EIAs and AEE for UK Spaceports

Part three provides data in (Table 1) on the status of EIAs for UK spaceports and the local planning authorities who are responsible for approving EIAs before an AEE, and license application is submitted to the CAA. The CAA as the UK regulator, is responsible for approving AEEs and license applications. To recap, the process (for environmental approval) of licensing space activities in the UK involves very local rules and procedures that expand outwards to encompass specific national and international governance instruments and guidelines.

1. The Local Planning Authority and EIA Process (Which Must Take Account of Local, National, and international environmental issues—from peat to marine areas to the climate).
2. environment is a devolved matter for Scotland, Wales and NI, EIA require their approval.

[50] Convention on Environmental Impact Assessment in a Transboundary Context (Espoo Convention) text of the Convention, 2017 (ECE/MP.EIA/21/Amend.1).

[51] Regulation 41, Town and Country Planning (Environmental Impact Assessment) (Scotland) Regulations 2017 https://www.legislation.gov.uk/ssi/2017/102/note/made.

Table 1 Spaceport EIA and AEE status *Source* created by author

Name of spaceport	Local planning authority (LPA)	Consultant(s)	Launch type	EIA status	AEE and license application
Submitted as "Shetland Space Center LTD" (Saxa Vord Spaceport)	Shetland Islands Council[52]	ITP energized	Vertical	Full EIA Submitted publicly available status: Pending	Not submitted
Space Hub Sutherland	Highland Council[53]	Ramboll environment	Vertical	Full EIA Submitted publicly available status: Permitted	Not submitted
Spaceport Cornwall	Cornwall Council[54]	University of Exeter	Horizontal	No EIA Emissions study[55]	Not submitted
Spaceport 1 North Uist	Council of the Western Isles[56]	Atkins SNC-Lavalin Group	Vertical	EIA 2nd step Scoping publicly available	Not submitted
Prestwick Spaceport	South Ayrshire Council (SAC)[57]	Unknown	Horizontal	EIA 1st step Screening (not public)	Not submitted
Snowdonia	Gwynedd Council[58]	Unknown	Horizontal	No EIA only a business plan[59]	Not submitted
Machrihanish	Argyll and Bute Council[60]	Unknown	Horizontal	No EIA	Not submitted

[52] Shetland Islands Council planning portal EIA for Shetland Space Centre Ltd. (renamed Saxa Vord Spaceport) 2021/005/PPF Vertical launch space port including launch pad complex, satellite tracking station, assembly and integration hangar buildings, with associated security fencing, access, servicing and infrastructure on Land at Lamba Ness, Unst, Shetland https://pa.shetland.gov.uk/online-applications/applicationDetails.do?activeTab=documents&keyVal=QMI5QYOAGFC00.

[53] Highland Council 20/00616/FUL Construction of vertical launch spaceport with launch operations control centre, site integration facility, launch pad complex, antenna park, access road, fencing, services and associated infrastructure on Land 2600 M SW Of Dunbuie Talmine Tongue https://wam.highland.gov.uk/wam/applicationDetails.do?activeTab=documents&keyVal=Q5CD2AIHKTF00.

[54] Author's Email Correspondence with Cornwall Council confirmed no EIA yet 8/16/21 Cornwall No EIA planning application or EIA available https://planning.cornwall.gov.uk/online-applications/simpleSearchResults.do?action=firstPage.

3. the final AEE must be accepted in the licensing application by national regulator the CAA.
4. the UK as part of ESPOO must engage in diplomacy / international cooperation on TEIA with other countries that may be affected by spaceport/launch activities originating in UK.

As shown, (Table 1) provides the most up-to-date publicly available information on the status of EIA and AEE for UK spaceports at the time of writing this chapter. While UK space regulations came into effect on 29th July 2021, as of 18th August 2021, the CAA has not yet received any application for any spaceport or launch license nor the required assessment of environmental effects.[61] The AEE's will be made available for public comment once they are received. Several of the spaceports in (Table 1) are planning to be operational by 2022–2023. Thus far, only Space Hub Sutherland has received approval from their local planning authority. Saxa Vord is still pending. Two others have begun screening and scoping.

EIAs for Space Hub Sutherland and Saxa Vord Spaceport address transboundary issues in their applications to their respective LPAs. For Sutherland (the only completed EIA with approval from their LPA the Highland Council) transboundary risks included the probability and impacts of launch vehicle stages and farings dropping in the marine environment of the Arctic thus falling within the jurisdiction of several countries including Iceland, Faroe Island (Denmark) and Norway (as well as several international agreements and monitoring programs). Similarly, for Saxa Vord Spaceport (EIA filed as Shetland Space Centre), prepared by ITP Energized, there will also be marine transboundary effects to contend with. Their EIA also reported that effects on the marine environment will arise from the return to Earth of launch vehicles and may occur in Scottish waters or in the waters of other countries also

[55] Xiaoyu Yan, *Spaceport Cornwall Carbon Impact Assessment* (University of Exeter 2019).

[56] Scoping documents 18/00234/SCO_L Spaceport 1 Project—infrastructure will comprise an admin/security building, two launch pad areas, two assembly buildings and fuel storage tanks and operations areas, as well as on-site access roads to connect these components to the main public road (the A865). Spaceport Facility Scolpaig Balemartin Isle Of North Uist https://planning.cne-siar.gov.uk/PublicAccess/applicationDetails.do?previousCaseType=Application&keyVal=PAM9RLRO02A00&previousCaseNumber=19%2F00311%2FPPD&activeTab=summary&previousKeyVal=PTPPUTRO02A00.

[57] Prestwick publicly stated they have filed a 1st step screening report but it has not been made public http://www.prestwickaerospace.com/prestwick-spaceport-files-environmental-screening-report/.

[58] https://amg.gwynedd.llyw.cymru/planning/index.html.

[59] No Snowdonia spaceport EIA found in the Gwynedd Council planning portal, further research found an earlier Airspace Change Proposal, ACP-2019–58, Llanbedr Danger Area (DA)—Public Consultation Easy Read Guide https://gov.wales/sites/default/files/publications/2021-01/atisn14685doc1.pdf and business plan https://ygc.cymru/wp-content/uploads/Llanbedr_Docs/A496_Llanbedr_Access_Improvement_SEZ_Economic_Impact_Assessment.pdf.

[60] No EIA planning documents for Machrihanish spaceport in the Argyll and Bute Council portal were found https://publicaccess.argyll-bute.gov.uk/online-applications/search.do?action=simple&searchType=Application.

[61] Author's Email correspondence with the UK Civil Aviation Authority and the author 17–18th August 2021.

specifically Denmark (Faroe Islands), Iceland, and Norway (including Jan Mayen). Due to these potential transboundary effects, Scottish Ministers must be consulted and may decide to send a copy of the EIA report to the governments of affected countries and invite them to participate in the consultation process.[62]

The Space Hub Sutherland EIA included TEIA performed by Ramboll Environment, a consulting firm known for wind farms and energy related EIAs:

> The Study Area is considered an important area for a number of ecological receptors, even though it has been largely influenced by human activities and associated pollution incidents. At the same time, the Study Area is considered to be characterized by a relative low density of human infrastructure and activities. Given the low number of anticipated launch events per year (up to 12) and anticipated ability of the receiving environment to cope with deposit of LV [launch vehicle] components generating debris (due to large open nature of the water body under consideration), overall risks for transboundary effects on water quality and ecological receptors are anticipated to be low and not significant. Similarly, assuming that some form of notice for mariners is issued prior to launch events, the overall risks for transboundary effects on humans and human activities are anticipated to be low and not significant. No further requirement for mitigation has been identified during the operational phase of the proposed development.[63]

The LPA Highland Council concluded that while there are transboundary effects, these are not significant in EIA terms and considered the following effects for marine space debris:

> "Marine debris impact of marine ecology – the planning authority consider that space debris poses a risk to marine life through direct collision and through ingestion of macro and microplastics. The planning authority note that as a result of the proposed 12 launches per annum a minimum of 5 tonnes of carbon fibre reinforced plastic and 7 tonnes of metal alloy will be dropped into the sea. The planning authority can not conclude from the information within the EIAR whether this is a significant increase on current level of debris but note that the risk is proportional to the number of launches. The planning authority intend to condition that no more than 12 launches can be undertaken per annum.[64]

The Highland Council decided that the transboundary effects are not significant in Space Hub Sutherland's EIA. Thus far, Highland Council has notified the UK Government (UK Space Agency) about Space Hub Sutherland's EIA and application and "they are in discussions on spaceports with the Icelandic, Faroese, Danish and Norwegian governments. They confirmed that there is no requirement to consult them [UK government] at this stage. They commented on how the risk of collision with mariners will be managed. They set out that The Space Industry Act 2018 establishes public safety as the primary duty of the regulator with regards to spaceflight."[65]

---

[62] Shetland Space Centre (Saxa Vord Spaceport) https://shetlandspacecentre.com/wp-content/uploads/2021/01/SSC-EIAR-V1-NTS.pdf.

[63] Space Hub Sutherland Environmental Impact Assessment Report: Volume 1: Non-Technical Summary February 2020 https://www.hie.co.uk/media/7788/environmental-impact-assessment.pdf.

[64] Shetland Space Centre (Saxa Vord Spaceport) https://shetlandspacecentre.com/wp-content/uploads/2021/01/SSC-EIAR-V1-NTS.pdf.

[65] Ibid. Full text of the MoU between UK and Iceland https://assets.publishing.service.gov.uk/government/uploads/system/uploads/attachment_data/file/1010867/UKSA_Iceland_MOU.pdf, and

4 Concluding Remarks

This chapter explored the environmental considerations required for licensing UK spaceports and launch activities. The first section provided a brief background on proposed spaceports and explained the role of assessment of environmental effects. Part two discussed the details of EIA and TEIA and how they apply. Part three presented data on the status of EIAs and AEE for UK spaceport licenses.

These concluding remarks will summarize findings, propose future research, as well as draw attention to applicable UN COPUOS Long-term Sustainability (LTS) guidelines and the need for more attention to EIA and TEIA as tools, already in practice legally and normatively, for realizing aspects of the long-term sustainability of space activities.

4.1 Findings

This research found that while the CAA and the UK Secretary of State will have the final say in licensing UK spaceports and launches, no single authority or organization is responsible for governing the environmental impact of space activities in the UK. In this regard, it is necessary to gain a deeper understanding of a complex governance regime that involves both space activities and protecting the environment. These governance duties, as explained in the previous sections are shared between the CAA, UK government, Scottish and Wales governments, several local planning authorities, international organizations, conventions, protocols, as well as input from a wide range of industry and other stakeholders including the concerned public. The UK's ambitions to build-up national spaceport and launch capabilities did not happen overnight. The discussions and planning at all levels of government

---

full text of the MOU and Demark (Faroe Islands) https://assets.publishing.service.gov.uk/government/uploads/system/uploads/attachment_data/file/914515/Faroe_Islands_Memoranda_of_understanding.pdf, See also an archive of emails and letter (some redacted) regarding Spaceport Related Planning Applications, TEIA decisions and MOUs with impacted countries a letter beginning on pg. 26 from Scottish Minister for Trade, Investment and Innovation reads "UK Government have recently signed a Memorandum of Understanding (MOU) with the Faroe Islands with the support of Denmark—and that UKSA is in discussion with Iceland on the same basis and working toward a similar objective. With regard to Norway, who as a country also hosting a spaceport are arguably our closest European competitor in the race to first European launch, will understand the potential impacts of launch activities. However, it is not anticipated that there will be any objection from the Norwegians. I hope this goes some way to illustrate that transboundary environmental matters related to UK launch have been of paramount importance to the Scottish and UK Governments. We are certain that discussions and cooperation in space programmes may represent an opportunity to strengthen our relations with our neighbours. I would be happy to meet with Nordic Governments to discuss further if required, and to maintain good relationships." https://www.gov.scot/binaries/content/documents/govscot/publications/foi-eir-release/2020/10/foi-202000070852/documents/foi-202000070852---information-released-b/foi-202000070852---information-released-b/govscot%3Adocument/FoI-202000070852%2B-%2BInformation%2Breleased%2B%2528b%2529.pdf

have spanned more than a decade and have continued throughout and post-Brexit. This has also not been a unilateral maneuver in the international realm, the UK has relied heavily on support from the United States government and commercial space industries and their expertise in spaceport and launch activities. The role environmental governance issues play in space activities is wide ranging and a substantial amount of information about planned space activities is found within the AEE-EIA process through publicly available documents. Finally, there has been a lot of media coverage about commercial space launch from the UK even though regulations and the regulator (CAA) only recently came into effect 29th July 2021. Thus far, no AEE or license applications have been received and it remains to be seen how quickly, once received, the applications will be approved.

4.2 Future Research

The limitations of what this chapter could cover also opens the opportunity to recommend avenues for future research. This chapter did not delve into public commentary nor the resistance to spaceports in various areas of Scotland and Wales. EIAs contain a wealth of information regarding views from environmental NGOs, other stakeholders, and the concerned public. This chapter also did not delve into the political aspects of EIA nor known weaknesses and flaws in EIA steps including the timing of when and how information is made available. Future research on EIA for space must contend with and account for the method's known political and methodological flaws. Suggestions for future research includes investigating the details of the diplomatic process involved in TEIA and subsequent MOUs that result in wider international cooperation but also risks to the receiving country as well as dealing with impacts to ABNJ. Another avenue for research, given the rising numbers of proposed spaceports around the world is investigating what comes from stakeholder and public pushback with regard to protecting the environment especially when these concerns are overridden by high-stakes national security interests (military, intelligence, and economic). Finally, more research should consider aspects of the UN COPUOS Long-term sustainability of space activities guidelines and a role for EIA & TEIA. Specific language in the LTS guidelines A.2 and D.1 advises signatories to protect the environments of both Earth and Space and could be interpreted as pointing to existing tools such as EIA and TEIA, especially D.1.4.[66]

[66] In 2019, the United Nations Committee on the Peaceful Uses of Outer Space (COPUOS), of which the UN Office for Outer Space Affairs (UNOOSA) is Secretariat, adopted the Guidelines for the Long-Term Sustainability of Outer Space Activities (LTS guidelines), full text of the guidelines A/74/20, para 163 and Annex II. 26 January 2021 The UK and UNOOSA announced an agreement to promote the LTS guidelines. The Promoting Space Sustainability Project, funded by the UK will inform future UNOOSA capacity-building efforts to promote the sustainability of outer space, and it will encourage all actors to implement the LTS guidelines to the fullest extent possible https://www.unoosa.org/oosa/en/ourwork/topics/promoting-space-sustainability.html.

Guideline A.2

2.In developing, revising, or amending, as necessary, national regulatory frameworks, States and international intergovernmental organizations should:

(c) Address, to the extent practicable, risks to people, property, public health, and the environment associated with the launch, in-orbit operation, and re-entry of space objects.

(d) Promote regulations and policies that support the idea of minimizing the impacts of human activities on Earth as well as on the outer space environment. They are encouraged to plan their activities based on the Sustainable Development Goals, their main national requirements and international considerations for the sustainability of space and the Earth.

Guideline D.1

3.States and international intergovernmental organizations should promote the development of technologies that minimize the environmental impact of manufacturing and launching space assets and that maximize the use of renewable resources and the reusability or repurposing of space assets to enhance the long-term sustainability of those activities.

4.States and international intergovernmental organizations should consider appropriate safety measures to protect the Earth and the space environment from harmful contamination, taking advantage of existing measures, practices and guidelines that may apply to those activities, and developing new measures as appropriate.

Nancy Riordan is a Ph.D. student in the Global Governance and Human Security program at the University of Massachusetts, Boston, USA. Her doctoral research focuses on global governance and space activities with an emphasis on EIA and TEIA as tools for realizing long-term sustainability goals.

Legal and Policy Dimension of UK Spaceports

Christopher Newman

Abstract From a virtual standing start at the turn of the 21st Century, the UK Government has increasingly looked upon space activity as a method of generating significant amounts of economic growth. In 2010, the Innovation and Growth Strategy recognised that the space sector had the potential to deliver sustained economic growth. Accordingly, the UK Space Agency (UKSA) was formed that same year, accompanied by sustained investment, including creating regional 'catapults' to commercialise the provision of data from space (The first satellite applications catapult was launched in 2013). This discussion will examine how space policy has shifted towards the reinstatement of a 'sovereign launch capability' within the UK, the legal and policy implications of this for the creation of UK spaceports. The analysis will cover the need for the UK to provide the necessary legal framework for the launch industry, discuss and critique that new framework, and consider some of the policy issues that are yet to be resolved.

1 UK Space Policy: Evolution and Revolution

The Government announced the first UK National space policy in 2015 and the National Space Strategy in 2021.[1] These initiatives have sought to maximise the return on investment from the space industry whilst recognising that space activity has strategic importance due to the infrastructure that depends on data from space.

[1] For the full policy, see https://assets.publishing.service.gov.uk/government/uploads/system/uploads/attachment_data/file/484865/NSP_-_Final.pdf accessed 25 August 2021. The National Space Strategy was promulgated in September 2021 and is available at https://assets.publishing.service.gov.uk/government/uploads/system/uploads/attachment_data/file/1020617/national-space-strategy.pdf.

C. Newman (✉)
Northumbria Law School, Northumbria University, Tyne, England
e-mail: christopher.newman@northumbria.ac.uk

A. Froehlich (ed.), *Spaceports in Europe*, Studies in Space Policy 34,
https://doi.org/10.1007/978-3-030-88311-9_6

Following the report on the health and state of the UK space industry,[2] the UK Government stated an ambition to grow the national share of that global space market to 10% by 2030.[3] All the activities of UKSA are still shaped by that core goal, and the National Space Strategy reiterates the value of space to the UK economy. In an era of economic retrenchment brought about by the global financial crisis, the UK has continued to increase its budgetary contribution to the European Space Agency (ESA).[4] The establishment in 2020 of the National Space Council (NSC) showed the commitment from the UK Government to integrate space within the very highest policy circles.[5]

Much of the work undertaken in recent years by the UK Government in developing the national space economy focused upon extant strengths within the UK space sector, such as the manufacture of small satellites. In 2017, however, the UKSA announced the UK Launch initiative[6] with the stated aim of being the first country in Europe to offer small satellite manufacturers a direct end to end route to launch, building on the success of the small satellite industry.[7] Three factors can be seen to be behind this shift in policy.[8] First, when the policy was being considered, there was a lack of availability of primary payload capacity. With this bottleneck, some small satellite customers were waiting 18 months before a suitable launch platform was available[9] Second, there appeared to be an increasing momentum around the world for states to develop a sovereign launch capacity.[10] Finally, and perhaps more cynically, it might also be viewed as an attempt by the UK Government to demonstrate the viability of

---

[2] http://londoneconomics.co.uk/blog/publication/the-case-for-space-2015/ accessed 25 August 2021.

[3] See above n2 at 7.

[4] Peggy Hollinger and George Parker, "UK increases funding to European Space Agency more than 15%" Financial Times, November 27 2019 available at https://www.ft.com/content/64926e9e-108c-11ea-a225-db2f231cfeae accessed 25 August 2021.

[5] The NSC is political body, made up of prominent politicians within the UK Government, and aims to provide strategic leadership and coordination on all aspects of space-related matters as they relate to government. See https://committees.parliament.uk/work/1194/uk-space-strategy-and-uk-satellite-infrastructure/ for more details. Accessed 25 August 2021.

[6] For further details see: https://assets.publishing.service.gov.uk/government/uploads/system/uploads/attachment_data/file/926377/LaunchUK_brochure.pdf accessed 25 August 2021.

[7] ibid at 2.

[8] Christopher Newman, "The Space Industry Act 2018: Unlocking the UK Space Economy" (2019) 70 Proc IISL 282–291.

[9] Stuart McIntyre "Growth drivers, requirements and threats in the smallest industry", (2018) 14(4) ROOM 34–38, at 35.

[10] A number of States have either developed or are in the process of developing launch capacity. See, for example, Foust, J., "Launch Canada" The Space Review 5 March 2018 available at http://www.thespacereview.com/article/3445/1 and Amos, J., "NZ rocket launch heralds new wave' BBC News, 22 January 2018 available at http://www.bbc.co.uk/news/science-environment- 42,780,872 accessed 25 August 2021.

'Global Britain' following the political and economic chaos surrounding Brexit and the UK's decision to leave the European Union.[11]

The irony of this is that in October 1971, the UK had become the sixth nation to independently launch a satellite into Earth orbit using a launch vehicle it had developed from a launch site in Woomera, Australia.[12] Unlike the other launch nations,[13] however, the UK decided to abandon this capability, citing the cost of development and alternative, cheaper arrangements with other states.[14] Nonetheless, by 2017, the UK government had formed the view that independent launch capacity was essential to further the ambition of capturing 10% of the global space market.[15] Therefore, the UK Launch initiative was announced alongside a £10million incentive scheme to try and develop commercial spaceflight from within the UK.[16]

2 The Evolution of UK Spaceport Policy

The creation of a spaceport, and the related infrastructure for commercial launch activity, is perhaps the most fundamentally important if a launch is to occur from within the British Isles. Despite UK Launch not emerging as policy until 2017, developing a UK spaceport has been something that has been considered within Government from as early as 2012.[17] In 2014, a national, state-sponsored competition was announced to select a site to build a commercial spaceport within the UK, funded by the Government. In July 2014, things had even gone so far as to see the commissioning of a report, looking across existing military and civilian airports to identify

[11] Newman, C.J., "The Precis: UK Domestic Space Law" Spacewatch Middle East, April 2017. Reproduced from Listner, M.J., The Precis, 31 March 2017 available at https://spacewatchme.com/2017/04/the-precis-uk-domestic-space-law/ accessed 25 August 2021.

[12] Christopher Newman, "Rediscovering UK Sovereign Launch Capability" (2018) 16(2) ROOM 89.

[13] The Soviet Union and the United States were already well established as space powers, but France, Japan and – most notably – China would go on to develop significant and enduring space programmes of their own, see for details Zhongyang, Zheng "The origins and development of China's manned spaceflight programme". (2018) 23 Space Policy 3 167–171.

[14] For details of the political, economic and diplomatic reasons behind the decision to cancel the Black Arrow project see Hill, C.N., "Black Arrow Cancellation", Space UK (Online) available at http://www.spaceuk.org/ba/blackarrowcancellation.htm accessed 25 August 2021.

[15] https://www.theguardian.com/science/2020/dec/31/is-the-uk-about-to-have-liftoff-in-the-global-space-industry accessed 25 August 2021.

[16] https://www.gov.uk/government/news/government-announces-boost-for-uk-commercial-space-sector accessed 25 August 2021.

[17] Parnell, B., "UK.gov to clear way for Britain's first Spaceport" The Register, 11 July 2012 available at https://www.theregister.co.uk/2012/07/11/uk_gov_open_skies_to_spaceplanes/ accessed 25 August 2021..

potential sites that could be repurposed for use as spaceports.[18] Initially, forty-six such places were identified.[19] Once non-operational and non-practical places (such as Heathrow or Gatwick, which were already over capacity) were discounted, a shortlist[20] of potential locations around the UK that would be suitable to host rocket launches had been drawn up.[21]

In May 2016, however, the Government abandoned the idea of a single, publicly funded spaceport. Instead, the UKSA announced that to 'avoid restricting the development of the UK (launch) market, the Government will create the regulatory conditions for any suitable location that wishes to become a spaceport to take the opportunity to develop and attract commercial space business.'[22] This decision indicated a significant shift in the way spaceports would be developed and, crucially, moved from a tendering process to a licensing framework.

The Space Growth Partnership was introduced in 2018 to target funding into areas that needed development, and explicitly, the vertical and horizontal launch facilities within the UK.[23] Smaller grants replaced large amounts of state support. These were intended to incentivise competing sites to provide a business case and secure the total funding commercially.[24] The introduction of financial incentives short of complete state funding has seen grants of nearly £40million provided to establish spaceports at various sites around the UK.[25] Nonetheless, the economic approach was only part of the equation. If the UK were to become an active launch nation with attendant infrastructure, it would need a radical overhaul of the legal framework governing the regulation of national space activity.

3 The Need for Regulatory Reform

When considering the regulatory framework for UK Launch, both domestic and international obligations need to be considered. As Smith and Leishman affirm,

---

[18] Government review of commercial spaceplane certification and operations: summary and conclusions, July 2014, at 53–54, https://www.gov.uk/government/publications/commercial-spaceplane-certification-and-operations-uk-government-review accessed 25 August 2021.

[19] For further details on this see Lesley-Jane Smith and Ruairidh Leishman, "Up, up and Away: An Update on the UK's Latest Plans for Space Activities" (2019) 44 Air & Space Law 1, 1–26 at 4.

[20] The five shortlisted sites were Campbeltown Airport, Glasgow Prestwick Airport, Stornoway Airport, Newquay Cornwall Airport and Llanbedr Airport.

[21] Amos, J., "UK Ministers issue spaceport shortlist" BBC News, 3 March 2015 available at http://www.bbc.co.uk/news/science-environment-31711083 accessed 25 August 2021.

[22] McArdle, H., "UK spaceport competition axed in favour of licensing model" The Herald, 20 May 2016 available at http://www.heraldscotland.com/news/14506625.display/ accessed 25 August 2021.

[23] https://www.ukspace.org/space-organisations/space-growth-partnership/ accessed 25 August 2021.

[24] https://www.aerosociety.com/news/the-magnificent-seven/ accessed 25 August 2021.

[25] Ibid.

"*access to launching infrastructure...compels a state to exercise a key regulatory function because, under international space treaty law, obligations are imposed on a launching state's government towards third states in the event of damage caused by their launches or by satellites in orbit*".[26] The UK is a signatory to four international treaties governing the responsibilities of states in respect of space activity.[27] Without legislative intervention, the UK regulatory framework in place before March 2018 would not have been adequately equipped to discharge its international obligations under these treaties, nor would the Government have been able to assure the safety or security of the launch of space objects from the UK. Accordingly, the was a need to reform the existing legislation, which had not been drafted to deal with launch from the UK.

Regarding the international treaty obligations that the UK needed to consider, the troika of Articles VI, VII and VIII of the Outer Space Treaty 1967 is most pertinent to a nation looking to develop launch capability. These Treaty Articles underpin the regulation of human space activity, and any launching state must ensure compliance. Article VIII deals with the requirements of registration and the requirements for States to register their space objects.[28] Article VI provides for the principle of state responsibility in outer space to the international community. If a nation-state or a non-governmental entity based within that nation-state does not adhere to the Treaty, the nation-state can be held internationally responsible.

Perhaps crucially for the provision of launch infrastructure, Article VII provides for national liability for the damage a launching state may cause to another state by means of their space object. This Article was given further effect by the Liability Convention of 1972. As a result, the notion of a 'launching state' was introduced into international law. A nation becomes a launching state if they are launching, procuring a launch, or allowing their territory and/or facility to be used for the purpose of a launch. As provided by the Liability Convention, Liability can be found in three different scenarios, either on the surface of the Earth, to aircraft in flight, or in outer space.

---

[26] Smith and Leishman n23 at 5.

[27] The UK has signed and ratified The 1967 Treaty on Principles Governing the Activities of States in the Exploration and Use of Outer Space, including the Moon and Other Celestial Bodies (Herein referred to as the Outer Space Treaty or OST). This was adopted by the General Assembly of the UN on 19 December 1966 by virtue of Resolution 2222 (XXI). It was opened for signature on 27 January 1967 and entered into force on 10 October 1967. The UK has also signed and ratified Agreement on the Rescue of Astronauts, the Return of Astronauts and the Return of Objects Launched into Outer Space (ARRA) was adopted by the General Assembly of the UN by virtue of Resolution 2345 (XXII)), and the Agreement entered into force in December 1968. The UK signed and ratified the 1972 Treaty on Convention on International Liability for Damage Caused by Space Objects (Herein referred to as the Liability Convention) which was adopted by the General Assembly of the UN by virtue of Resolution 2777 (XXVI) and entered into force on 1 September 1972.Finally, the UK signed and ratified the Convention on Registration of Objects Launched into Outer Space which was adopted by the General Assembly in 1974 (General Assembly resolution 3235 (XXIX)) and entered into force on 15 September 1976 (Herein referred to as the Registration Convention).

[28] This is further expanded upon in Registration Convention.

By engaging in the activity of launching a space object, the UK would become a launching state. Consequently, it would attract liability for damage caused by space objects to anything on the surface of the Earth, to aircraft in flight and damage caused to the space objects of other states if the UK was shown to be at fault.[29] For damage caused on the surface of the Earth or to aircraft flight, liability is on an absolute liability basis (Art II Liability Convention). Still, this liability applies to other States, and those within the UK would not enjoy this protection. Therefore, there needed to be a regulatory framework that not only puts in place a safety-focused licensing regime for the operation of a spaceport, but also provision within domestic legislation to provide the equivalent protection for UK nationals for damage caused by launch as is available to other nations under Art II of the Liability Convention.

It has been recognised that the provision of independent, sovereign launch capability is a significant factor in any nation's ability to expand its national space industry.[30] The small, agile start-up companies, fundamental to the stated policy goal of capturing 10% of the global space market by 2030,[31] would either have had to look abroad for launching or abandon their plans altogether. Had the regulatory regime remained untouched, launch from the UK would not have been possible, and companies may well not have had either the money or other resources to fulfil the licensing requirements. Therefore, this discussion will examine the legislative and regulatory approach adopted by the UK Government necessary to promote the operation of spaceports within the UK.

4 The Space Industry Act 2018 and Spaceport Regulation

Before the coming into force of the Space Industry Act 2018, the legal basis for all space activity carried out by the United Kingdom was to be found in the Outer Space Act 1986.[32] This piece of primary legislation looks to discharge the obligations of the UK under the Outer Space Treaty 1967. The 1986 Act addresses the specific duty under Article VI of the Treaty regarding the authorising space activity of non-governmental entities under its jurisdiction. The 1986 Act confers the power to licence space activities upon the Secretary of State. However, the Act has been criticised by both academic observers and those in the UK space industry for several reasons. First, the licensing process itself is 'high cost' and burdensome. Also, there is a liability cap requiring the operator to indemnify the Government for loss or

[29] See Art III of the Liability Convention. Specifically, for damage caused elsewhere other than the surface of the Earth or to an aircraft in flight (i.e., for all operations in outer space), Art III Liability Convention provides that a state will be liable for damage caused to the space object of another state on a fault basis.

[30] Smith and Leishman, above n23 at 5.

[31] Above n2.

[32] The Outer Space Act 1986 is *still* in force and deals with the regulation and authorisation of UK space activity outside UK territory.

damage up to a limit of €60 million (which is seen as prohibitively expensive for smaller satellite operators).[33]

Crucially for UK Launch ambitions, the 1986 Act does not have express provisions for the authorisation of launch from the UK and managing the obligations outlined above imposed by the international treaties upon States that launch objects into outer space. Recognising the need for an appropriate legal framework for the licensing of spaceports and the conduct of space activities, including a launch from UK soil, the UK Government passed the Space Industry Act 2018. This Act was pushed through the UK Parliament by Government ministers unusually quickly to reassure potential investors in spaceports.[34] Such speed meant that much of the Act was skeletal and left much of the granular details to be decided utilising secondary legislation and regulations. Nevertheless, the Act represents the legislative phase of the UK seeking to regain an independent launch capability and puts in place a licensing framework for space activity from within the UK. Indeed, the 2018 Act goes further as it attempts to regulate the establishment of several small satellite launch facilities within the UK to complement its burgeoning small satellite industry.[35]

This new regulatory regime is intended to work alongside the UK Launch policy, which outlined a three-step ambition for embedding sovereign launch capacity within the UK. As has already been identified, the goal is to contribute to the UK capturing 10% of the share of the global space market by establishing the UK as a successful, reliable, and sustainable commercial gateway to space with integrated supply chains. Whilst the 2018 Act contains statutory authority for regulating spaceflight activities within the UK; there was a lot of detail that needed to be furnished. The passing of the 2018 Act provided the statutory authority for creating a detailed regulatory regime for the vertical and horizontal launch of objects into space. In early 2020, the UKSA announced the promulgation of the draft regulations to accompany the Space Industry Act 2018. The consultation ran over the summer of 2020 and obtained operational and stakeholder input into the underpinning secondary legislation.

There were three sets of regulations that were opened for consultation by the UKSA in 2020. The Space Industry Regulations, accident investigation regulations[36] and regulations managing appeals against decisions made under the two licensing regimes.[37] The consultation results were disseminated on 5th March 2021, and the amended regulations came into force on 27th May 2021.[38] Further discussions on insurance and environmental issues are ongoing as the Government seeks to develop

[33] Christopher Newman, and Michael Listner "A Very British Coup: Lessons from the draft UK Regulations for Cubesats." The Space Review, 31 August 2015, available at http://www.thespacereview.com/article/2816/1 accessed 25 August 2021.

[34] Newman, above n16.

[35] The Space Industry Bill received Royal Assent on 15 March 2018 after a relatively swift passage through Parliament. Hereafter the Space Industry Act will be referred to as the SIA.

[36] The Spaceflight Activities (Investigation of Spaceflight Accidents) Regulations 2021.

[37] The Space Industry (Appeals) Regulations 2021.

[38] https://www.legislation.gov.uk/ukdsi/2021/9780348223682/contents accessed 25 August 2021.

stakeholder involvement and ownership of the ultimate regulatory framework.[39] The liability question for space activity granted to other States under Art VII of the Outer Space Treaty is dealt with under s34 of the 2018 Act. The protection for damage caused on the surface of the Earth or to aircraft in flight within the UK is on the basis of strict liability (as opposed to the absolute liability mentioned in the Treaty). Whilst not directly relating to activities on a spaceport, resolution of the insurance requirements on spacecraft operators will undoubtedly be crucial to the success of spaceports.

5 The Space Industry Regulations and the Spaceport Licence

Part 2 of the Regulations provide that the Civil Aviation Authority (CAA) is the sole regulator for commercial spaceflight activities regulated under the 2018 Act.[40] Part of the reason for this is to remove the longstanding difficulty of having the UK Space Agency (UKSA) as the lead agency promoting UK space activity whilst also acting as a regulator for that activity. Additionally, the segregation of airspace for both vertical and horizontal launch will be a crucial issue for spaceport operators, so it makes sense to have a regulatory authority responsible for the use of airspace also accountable for regulating launch. Finally, the CAA is focused on safety. A principal focus of the new regulatory regime is to ensure that risk the public from spaceflight operations and the operation of a spaceport is as low as practicable. For this reason, the CAA represents a sensible and pragmatic choice as the regulator.

The UK, therefore, has two separate statutory licensing regimes for space activity *depending on where the launch occurs.* If the launch is procured outside the UK or satellite operations are conducted from facilities outside the UK, the operators will require a licence under the Outer Space Act 1986. On the other hand, the creation and operation of a spaceport together with range control, the launch (both horizontal and vertical, orbital and sub-orbital) and the operation of a satellite from within the UK (referred to as 'spaceflight activity in Part 1 of the 2021 Regulations) will be authorised and licensed under the Space Industry Act 2018.

Part 3 of the Regulations provides three distinct categories of licence that will be made available under the 2018 Act: an operator licence, a range control licence, and a spaceport licence. The *Operator Licence* is covered in Part 4 of the Regulations. This will possibly be the most requested licence and applies to those looking to engage in launch, re-entry, and satellite operations under the 2018 Act. *Range Control Licences,*

[39] Department for Business Energy, Innovation and Skills (BEIS) (2021) Unlocking Commercial Spaceflight for the UK: Space Industry Regulations Consultations: summary of views received and the Government's response. Available at: https://assets.publishing.service.gov.uk/government/uploads/system/uploads/attachment_data/file/968535/Space-Industry-Regulations-Consultations-summary-of-views-and-government-response-accessible.pdf at 7 accessed 25 August 2021.

[40] Space Industry Regulations 2021, SI 2021/792 Reg 3.

which are essential for the safe operation of spaceports, are covered by Part 6 of the Regulations and regulate range control services for activities licensed under operator licenses. Finally, *Spaceport Licences* are found in Part 5 of the 2021 Regulations. This covers all aspects of the operation of a spaceport within the UK and imposes distinct requirements for horizontal and vertical launching and detailing the need for an environmental impact assessment before issuing a licence.

As stated above, the development of spaceports has, since 2016, been viewed by the Government as a private-sector endeavour. The 2018 Act provides the bare bones of the approval process for spaceport operators and range safety operations. However, it is recognised that the desire for launch infrastructure needs to be balanced against safety and local environmental considerations. In granting a licence to operate a spaceport, the regulator must consider the general licence requirements under s.8 of the 2018 Act, the s.10(a) requirement that the regulator must not grant a spaceport licence unless satisfied that the applicant has taken all reasonable steps to ensure that risks to public safety arising from the operation of the spaceport are as low as reasonably practicable (ALARP), and the submission of an assessment of environmental effects required under s.11 of the 2018 Act.

The 2021 Regulations augment these basic statutory requirements. In the notes for guidance to the Regulations,[41] the underpinning ethos behind the regulation is to enable the regulator to ensure that the proposed location of the spaceport is appropriate for the type of spaceflight and associated activities and that the risks to public safety from the activities at the spaceport are reduced to a level that is ALARP. Additionally, the Regulations require the spaceport licensee to appoint named individuals to act as *accountable manager* (responsible for the effective financial management of the spaceport[42]), a *safety manager* (responsible for the day-to-day administration and maintenance of an effective safety management system[43]) and a *security manager* (responsible for all aspects of spaceport security).[44]

The approval and safety provisions concerning spaceports are covered extensively in Part 10 of the 2021 Regulations. Reg 152 imposes a duty upon spaceport licensees to ensure that the activities carried out on the spaceport are safe and requires licensed activities to be carried out in accordance with the current safety case by preventing a major accident from occurring[45] and mitigating the consequences of such an accident if it does happen.[46] With the requirement of emergency response planning under Reg 165. Understandably, the response to the consultation recognised these provisions as a welcome and sensible addition to the skeletal infrastructure. Given the proximity

[41] Available at https://www.legislation.gov.uk/uksi/2021/792/pdfs/uksiem_20210792_en.pdf accessed 25 August 2021.

[42] Space Industry Regulations 2021, SI 2021/792 Reg 7(2).

[43] Space Industry Regulations 2021, SI 2021/792 Reg 163 requires all Spaceports to have an effective safety management system. The matters to be addressed by this system are provided for in Schedule 4 of the Regulations.

[44] The responsibilities of a security manager can be found in Space Industry Regulations 2021, SI 2021/792 Reg 169.

[45] Space Industry Regulations 2021, SI 2021/792 Reg 152 (2) (a)(i).

[46] Space Industry Regulations 2021, SI 2021/792 Reg 152 (2) (a)(ii).

of the northern launch sites to the sea and the inherently dangerous nature of launch, the involvement of additional specialist environmental and maritime conservation experts may well be required as the regulatory regime for spaceports develops.[47]

The UK Government is keen to enable the licensing of launch activities from a ship,[48] subject to other conditions being met, and will provide further guidance on this. Neither the 2018 Act nor the 2021 Regulations recognise a ship as a spaceport. Launch from a ship would not require a spaceport licence but will still be regulated by a launch operator licence. The actual launch from a ship is within the scope of a 'spaceflight activity' as defined in the 2018 Act, whereas a ship carrying a launch vehicle to sea to launch is not.[49]

The onerous question of security at spaceports is dealt with in Part 11 of the Regulations. This details a whole range of security requirements, physical and personnel, cyber, training and vetting. Additionally, under regulations 192–202, there is a recognition of the need for special provisions relating to the protection of US technology. The UKSA has recognised that this requirement is needed to discharge the UK's responsibility under the US-UK Technology Safeguards Agreement signed in October 2020.[50] Therefore, despite some concerns being expressed during the consultation, the Regulations cannot be amended without compromising the UK treaty obligations and the involvement of US firms in the launch process.

6 Vertical and Horizontal Launch Within the UK

With the coming into force of the Space Industry Regulations 2021 to supplement the Space Industry Act 2018, the UK now has a licensing framework that can discharge the international obligations under the international space treaties. It also puts the onus on regulators to ensure that the risk to the surrounding area and any inhabitants are ALARP. Nonetheless, questions remain over the commercial viability of UK launch industries. Competition within the global launch industry remains fierce. Gwynne Shotwell, COO of SpaceX, speaking in 2019, stated that it is unlikely all the recent entrants into the launch market will survive.[51] With the USA boasting an array of established launch suppliers, costs are being driven down. The global market is also increasing in the range of nations and companies offering launch services.

---

[47] Space Industry Act 2018, s11(2) already requires a spaceport operator to submit an environmental impact assessment alongside the application for a spaceport licence.

[48] See for example the work being undertaken by Black Arrow Space and the support it is receiving from the UKSA https://blackarrow-space.uk accessed 25 August 2021.

[49] Space Industry Regulations 2021, SI 2021/792 Reg 15.

[50] For further details on this see: https://www.gov.uk/government/publications/ukusa-agreement-in-the-form-of-an-exchange-of-notes-between-the-united-kingdom-and-the-united-states-of-america-on-technology-safeguards-associated/uk-us-technology-safeguards-agreement-tsa-for-spaceflight-activities-understanding-the-tsa accessed 25 August 2021.

[51] http://interactive.satellitetoday.com/via/july-2019/how-launch-providers-are-standing-out-in-a-new-commercial-market/ accessed 25 August 2021.

The UK currently has seven sites vying to be spaceports.[52] It is suggested that, given current global trends, it is unlikely that the UK launch industry would be able to support seven spaceports given the costs associated with building the necessary infrastructure.

Therefore, the Government is looking to encourage horizontal and vertical launch facilities at spaceports across the UK to maximise opportunities for launching smaller satellites. There is no indication that the UK is looking to launch larger rockets at this stage. So, spaceports will be looking to attract customers from the small satellite market in Low Earth Orbit (LEO). For smaller satellites and those looking to go into polar orbit, the northern latitude of the UK is advantageous, being also as it is surrounded by sea, launches can be away from populated areas.[53] So, for example, launches from the Shetland Space Centre will launch in a northerly direction to avoid North Sea Oilfields and the Faroe Islands and will look to undertake up to 30 launches a year when fully operational.[54] The other vertical launch sites at Sutherland and Uist in the Hebrides both aim to provide regular vertical launches utilising this geographical advantage.

As the title suggests, horizontal launches involve an aircraft taking off from a runway, using a modified Boeing 747 to launch a modified rocket carrying a small payload. Regulations provide that a horizontal spaceport must be located at a certified aerodrome.[55] Spaceport Cornwall has already signed a partnering agreement with Virgin Orbit, a company that has already successfully conducted launch activity. Their modified 747 carried the rocket 'LauncherOne' up to an altitude of 35,000 ft, which then was deployed and reached orbit, from which ten CubeSats were then launched.[56] The Cornish spaceport is hopeful of flying the first Virgin Orbit flight in 2022. On a smaller but no less impressive scale, Snowdonia Aerospace is exploring the utility of using low-cost stratospheric balloons, augmented by small rockets to deploy very small satellites.[57]

---

[52] These are; Spaceport Cornwall in Newquay Cornwall, Glasgow Prestwick Airport, Campbeltown Airport in Kintyre, Snowdonia Aerospace in Llanbedr Wales (all horizontal launch) and Spaceport 1 – North Uist in Hebrides, Shetland Space Centre, Unst in the Shetland Islands and Space Hub Sutherland, Melness in Sutherland (which are vertical launch facilities).

[53] ibid.

[54] See https://shetlandspacecentre.com for details accessed 25 August 2021.

[55] Space Industry Regulations 2021, SI 2021/792 Reg 35 Aerodromes Regulation (Commission Regulation (EU) No. 139/2014 of 12 February 2014 laying down requirements and administrative procedures related to aerodromes).

[56] Christian Davenport "Virgin Orbit rocket reaches Earth orbit, adding an entrant to the commercial space race". The Washington Post 17 January 2021 available online at https://www.washingtonpost.com/technology/2021/01/17/richard-branson-virgin-orbit-launch-success/ accessed 25 August 2021.

[57] https://www.insidermedia.com/news/wales/newport-company-launches-balloon-to-demonstrate-spaceport-feasibility accessed 25 August 2021.

Each of the seven UK spaceports boasts unique characteristics that make them potentially attractive. They all provide significant employment opportunities within the local communities, many of which have seen a decline in traditional manufacturing or oil and gas industries. This is offset, however, by environmental concerns that exist amongst these communities.[58] Additionally, the UK has a significant amount of air traffic that flies over it. A critical practical difficulty facing spaceports is managing air traffic and the need for segregated airspace. As previously stated, the CAA is a regulator not only for UK spaceports but also has overall responsibility for determining the policy for the use of airspace. The CAA will then have the advantage of having an overview of the impact of a launch, not just in respect of safety on the ground but also to aircraft and the commercial interests of air carriers. This is a balancing act that will only become more difficult should the launch market in the UK realise its potential.

7 Conclusions

The decision of the UK Government in 2017 to prioritise the development of an independent, sovereign launch capability was understandable, given the desire to grow the space industry within the UK. Yet, the development of infrastructure from a virtual standing start means that funding and expertise are at a premium in the development of UK spaceports. The choice of locations makes sense geographically. It capitalises on the northern latitude and low population densities. Politically it is also advantageous as spaceports could generate significant economic growth in rural areas.

Nonetheless, while environmental concerns persist, it is the crucial economic argument that will ultimately determine the success and longevity of UK spaceports. The global launch market is more competitive than ever, and other nations are looking to develop launch capability. As a result, UK spaceports need to offer launch services that are competitively priced while recouping investment for costly infrastructure. It is not an easy task that lies ahead for UK spaceport operators.

Looking towards the regulatory framework, the UK Government has put a legislative framework in place with the avowed aim of enabling a competitive UK launch industry and satisfying the UK's international treaty obligations. The regulatory framework will ultimately be judged upon how it enables spaceport operators to engage in safe and profitable activity while protecting public safety. How liability and insurance will be dealt with under the 2018 Act is still yet to be determined. This, again, represents something of a balancing act. The Government is understandably reluctant to inhibit small and medium space companies by requiring expensive indemnity insurance. Yet, it must also ensure a robust framework to protect the wider

[58] https://envirotecmagazine.com/2020/02/21/planned-uk-spaceports-could-accelerate-global-warming/ and https://www.theguardian.com/environment/2014/sep/27/snowdonia-national-park-uk-spaceport accessed 25 August 2021.

public from damage and limit the liability to the public purse. This balancing act, coupled with the competition in the global launch market, will ultimately determine the success or otherwise of UK spaceports.

Dr. Christopher Newman is Professor of Space Law and Policy at Northumbria University. He is a member of the International Institute of Space Law and has presented his research to the International Institute of Space Law symposium on the regulation of UK spaceflight. He has also addressed the Committee on Space Research (COSPAR) planetary exploration committee regarding the legal implications of the discovery of non-terrestrial life. He has been an observer member of the UK delegation to the United Nations Committee on the Peaceful Uses of Outer Space (UN COPUOS). He is a Visiting Professor in Space Law at the Open University and an academic and policy consultant to Northern Space and Security, a space situational awareness and tracking company.

Spaceport: The Bridge Between Aviation Law and Space Law. The Grottaglie Sample

Maura Zara

Abstract In October 2020, the Italian Civil Aviation Authority (ENAC) has approved the *Regulation on Construction and the Operations of Spaceports*. Airport Grottaglie, in the province of Taranto, has now officially become the first Italian spaceport. This article aims, through the analysis of the regulation, to detect the most delicate legal details related to the establishment of such structures, as well as the particularly vivid interdependence of space law and air law. This will be revised in specific relation to the services of spaceports, where the study aims to find the basic legal structure in which the intrinsic peculiarities of space activities can be grafted. It will then proceed with the identification of national and international legislation referred to in the document, to continue with the essential issues that the establishment of a spaceport involves, as well as the need to ensure a sustainable use of the territory where the structure is located and the prevention and management of the risks inherent in the activities is carried out. Particular attention will be given to the historical relationship of continuity that air and space law have in the specific perspective of the establishment and management of spaceports, bringing the Italian example of Grottaglie. Now that Italy sees the birth of its first spaceport, catapulting itself towards a future already present, the analysis will conclude with a reflection on the challenges and benefits of the project from a national and international point of view.

1 Introduction

In October 2020, the Italian Civil Aviation Authority (ENAC) has approved the *Regulation on Construction and the Operations of Spaceports*. Airport *Marcello Arlotta* of Grottaglie, in the province of Taranto, has now officially become the first Italian spaceport. *Aeroporti di Puglia* is the company that manages the port of call.

According to what prescribed from the Italian Minister of Infrastructures and the Transports with Decree n. 250 of 2019, thanks to the adoption of the Regulations, it

M. Zara (✉)
Space Law Researcher at First Steps Legal, London, UK
e-mail: maura@firststepslegal.co.uk

A. Froehlich (ed.), *Spaceports in Europe*, Studies in Space Policy 34,
https://doi.org/10.1007/978-3-030-88311-9_7

can begin the process that will allow to use the airport of Taranto-Grottaglie also as national spaceport. Italy is thus the first European country to adopt a Regulation to implement and manage a spaceport on its national territory.

The drafting of the Document (to which the most important public and private actors of the sector have contributed) constitutes an important implementation of government strategies in the field of Space Economy; and concerns, in particular, the regulation of sub-orbital commercial activities, in fact, when the project was undertaken, reference was made essentially to sub-orbital flights.

Grottaglie is located in an area that has climatic conditions that allow throughout the year the testing activities; in addition to the favourable technical-formal aspects, it is also a site that has attracted projects related to the space economy (for example drones, hyperdrones, satellites and today suborbital flights of people and goods).

Before proceeding to the discussion, it is appropriate to clarify what is a spaceport. As stated in the Definitions section, paragraph three of the Document:

> Spaceport is a site, identified within the area of an aerodrome certified under the Regulation (EU) no. 139/2014, whose infrastructural and facilities equipment as well as its technical requirements allow horizontal launch/take-off, re-entry/landing and ground/flight operation of a single-stage or multistage reusable suborbital vehicle. The site is structured to allow all the necessary operations to execute a suborbital flight, including security controls for access of people and goods into security restricted areas, suborbital vehicle and the related systems maintenance and preparation to flight, training of crew and other occupants.

It is therefore an infrastructure that allows suborbital transport operations to be carried out, where experimental suborbital flights can be conducted (for example, microgravity testing, research and technological development), as well as operations for the training of highly specialized personnel (specifically, training for astronauts and crew).

The site includes buildings, installations and equipment to allow the launch, re-entry, related ground and airborne operations of a horizontal take-off and landing suborbital vehicle. Moreover, in the future, further purposes could be identified, as well as the installation for operations of vertical launch and landing (Vertical Spaceport), which will require the approval of ENAC.

The Document contains various regulatory innovations, in particular with regard to the safety of operations and the protection of the territory. Its analysis will demonstrate also the close interdependence between space law and air law, as well as the international dimension in which the regulation of these structures necessarily takes root, as will be deduced from the identification of national and international Legislation referred to in the document.

2 The Historical Relationship of Continuity Between Air and Space Law

As is well known, the beginning of the Space Law era is generally traced back to 1957, when the first artificial satellite in outer space was launched by the Soviet Union.

However, the reflection of jurists on the matter began well before that historic event,[1] and it has found fertile ground in air law. Therefore, the interdependence of air law and space law goes back to the very birth of this new branch of international law.

In fact, the first reference to space law dates from to 1910, in a brief comment on air law. The author was the Belgian jurist Emile Laude,[2] who brought attention to the fact that the atmospheric and the extra-atmospheric environment, being different in substance, required different forms of regulation.

We must wait until 1926 to find a further reference to the matter, again focused on air law. During a conference held in Moscow in December of that year, a senior official of the Soviet Ministry of Aviation, V. A. Zarzar, presented an article on the subject of limits to extension of national sovereignty in the atmospheric space. However, he went much further, introducing the argument of interplanetary flights.[3]

It can be stated that space law has found in air law the basic legal structure in which the intrinsic peculiarities of space activities can be grafted. In particular, the spaceport entity presents itself as a junction point between civil and space aviation, showing the ease with the two disciplines meet themselves and advance towards the new opportunities that the aerospace sector offers.

Paradigmatic is the choice to implement an airport facility with the development of infrastructure that allow access to outer space. Evidently, it's not a random choice.

The relationship of continuity between air and space law in the sector of aviation and transport activities could not be more evident, in a mutual interchange relationship that allows the development of both disciplines and the related activities under them.

It is the Document itself that shows how the project for the establishment of the spaceport is embedded in a legislative basis that is deeply rooted in aviation legislation, as it will be possible to see in the following paragraph. This is because it is the perfect framework, sharing many of the needs that the establishment of a spaceport entails.

Of course, it should be noted that the peculiarities of space activities, and in particular of extra-atmospheric flights, involve a necessary adaptation and the development of legislation that meets the precise requirements of operations carried out within a spaceport.

---

[1] For further information on the first references on space law, see Vol. 1 Doyle, S. E., Origins of International Space Law and the International Institute of Space Law of the International Astronautical Federation, International Institute of Space Law, Univelt, 2002.

[2] Laude, E., Question Pratiques, in Revenue Juridique Internationale del Locomotion Arienne, Vol. 1, Paris, 1910, 16–18. Cito da Doyle, S. E., A Concise History of Space Law: 1910–2009, in New Perspectives on Space Law of the 53rd IISL Colloquium on the Law of Outer Space, Yang Scholars Session, 2011, 1–24.

[3] Zarzar, V. A., Mezhdunarodnoye Publichnoye Vozdushnoye Pravo, in Voprosy Vozdushnogo Prava, Sbornik Trudov Sektsii Vozdushnogo Prava Soyaza Aviyakhim (Soyuz Obshchestv Druzhey Aviatsionnoy i Khimicheskoy Oborony i Promyshlennosti), Vol. 1, SSSR i Aviakhim RSFSR, Moscow, 1927, 90–103.

3 Identification of National and International Legislation Referred to in the Document[4]

In order to highlight the continuity of the two branches of aviation and space law, it should be analysed the text of the Regulation, starting from the identification of the national legislation referred to in it.

In the References section, the document comprises, emblematically, the Italian Navigation Code, Part II—Air Navigation; Decree 13 September 2002, no. 263 "Regulation on leasing of airport security services, modification of the ministerial decree 29 January 1999, no. 85, bringing the implementation rule of the article 5 of the Decree-law 18 January 1992, no.9, converted, with modifications, by the law 28 February 1992, no.217" and the National Programme for Civil Aviation Safety and the National Programme for safety quality control in civil aviation.

It is then extremely interesting to note the fundamental contribution of European legislation, which has been widely referred to in the field of civil aviation: Regulation (EC) no. 300/2008 of 11 March 2008 and following modifications and integrations, that establishes shared rules on civil aviation safety and that abrogates the Regulation (EC) no 2320/2002; Regulation (EU) no. 1998/2015 (and related Decisions) of 5 November 2015 and following modifications and integrations that establishes detailed instructions for the implementation of the fundamental shared rules on air safety, as modified by the Regulation (EU) 2020/111 of 13 of January 2020 "that modifies the implementing regulation (EU) 2015/1998 regarding safety equipment of the civil aviation and the third countries which are believed to apply safety rules that are equivalent to the fundamental shared rules on civil aviation safety" and Regulation (EU) no. 1254/2009 of 18 December 2009 that defines the criteria to allow the Member States to avoid the fundamental shared rules for civil aviation safety and to adopt alternative safety measures.

Finally, the Regulation mentions the Doc. ICAO 9137-AN 898 "Airport Services Manual" Part 1 - Rescue and Firefighting.

Therefore, it is clear that space law has a debt with air law, where it finds the guidelines to graft the regulation that meets its specific needs. In fact, as already mentioned, the specificities of sub-orbital transport operations require a legislation specifically related to these activities. The Document mentions: ED Decision 2016/027/R EASA: Certification Specifications and Guidance Material for Aerodromes Design (CS-ADR-DSN) and ED Decision 2014/012/R EASA: Acceptable Means of Compliance (AMC) and Guidance Material (GM) to Authority, Organisation and Operations Requirements for Aerodromes (AMC – GM).

It should also be said that ENAC has enacted the Regulations in coherence with the act 10 July 2017 n. 354 of the Minister of Infrastructure and Transport on the sustainable development of the sector of commercial suborbital flights.

Further national and international documents, referred to in the document, are specifically related to the issue of security and responsibility for the activities carried

[4] Section A, Reference, page 7, Regulation on Construction and Operations of Spaceports.

out in spaceport: Legislative Decree 26 June 2015 no. 105 "Implementation of the directive 2012/18/EU on the control of major-accident hazards involving dangerous substances"; Decree of the Minister of the Interior 7 August 2012 "Instructions on how to submit an application in relation to the procedures to prevent fires and the documentation to attach, under the Art. 2, clause 7 of the decree of the President of the Republic 1 August 2011, no. 151"; Decree 3 August 2015 "Approval of the technical rules to prevent fires, under the art. 15 of the Legislative Decree 8 March 2006, no. 139" and Decree of the Minister of the Interior 9 May 2007 "Directives for the implementation of the engineering approach to fire safety".

4 The Essential Issues that the Establishment of a Spaceport Involves

The installation of a spaceport, as any other industrial operation involving high-risk activities, means the use of special precautions. From the safeguard of the territory where the structure rises and the activities are carried out, to the prevention of the risks and potential disasters that can derive from them; as well as the particular attention to the perpetration of unlawful acts in the structure operations.

It should be noted that these sensitive issues find a regulatory network dictated for airports that is perfectly suited to spaceports, with the appropriate adjustments.

4.1 The Need to Ensure a Sustainable Use of the Territory Where the Structure is Located

A significant concern in the drafting of the document was the need to ensure sustainable use of the territory where the spaceport is located:

> In order to avoid improper use of land and to make the best use of the infrastructural resources present, the spaceport is identified in the area of an aerodrome certified under the Regulation (EU) no. 139/2014 and owning an ICAO code of flight infrastructure suitable for the characteristics of the suborbital vehicle that will operate there. The use of the main infrastructures, primarily the runway, will be shared and, normally, it will not be simultaneous.[5]

It is therefore a primary objective to respect the territory, avoiding its improper deterioration, making the best use of the infrastructure resources, which are adapted to the suborbital vehicle characteristics. This is why the spaceport is located at a certified airport.

We can also observe that for the reduction of the suborbital vehicle impact on wildlife, the measures already in place for airports has extended to the spaceport.[6]

---

[5] Introduction, par. 3, Regulation on Construction and Operations of Spaceports.

[6] Section B, Chap. 7, par. 8, Regulation on Construction and Operations of Spaceports.

Moreover, the protection of the territory from the risk of accident generated by the flight activity is also implemented through the Risk Plans for the spaceport. The Regulation defines Risk Plan:

> Document that includes the provisions and limitations to be transposed in the urban planning of each Municipality where the spaceport area is located, pursuant the provisions of the 5th amendment of the Art. 707 of the Air Navigation Code aimed at preserving the territory from the consequences of a potential accident or serious incident.[7]

This document is drawn up by the municipalities, where territory, the spaceport and the associated flight corridors are located, on the basis of the flight corridors[8] provided by the spaceport manager. The Spaceport operator[9] must provide the municipalities with a map of the flight corridors approved, together with the geographical coordinates to detect them, according to a method accepted by ENAC.[10]

4.2 *The Prevention and Management of the Risks*

Activities related to the exercise of a spaceport involve intricate risks,[11] being able to determine critical events with harmful effects on people, goods and the environment. Such risks are often foreseeable and can be adequately managed by appropriate protocols. A plan for emergency management is therefore essential to eliminate or at least mitigate the effects in a timely manner.

Consider, for example, the risks associated with the operations of the suborbital vehicle, the possibility of unlawful acts, sabotage, the danger of fires or explosions.

The Regulation expressly provides that:

> The spaceport operator is obliged to immediately notify ENAC of events connected to suborbital operations that may risk to endanger persons, goods and environment related to the spaceport, aerodrome and surrounding areas. After the event took place, the spaceport operator is responsible for taking the necessary measures to ensure that the spaceport (or aerodrome) areas and flight corridors exposed in terms of safety point of view are not used by aircrafts and suborbital vehicles.[12]

The fundamental importance of the analysis of the risks associated with the activities carried out emerges from the fact that it is the basis for obtaining the spaceport

---

[7] Section B, Chap. 7, par. 6, Regulation on Construction and Operations of Spaceports.

[8] The document defines Flight corridor an area of the Earth's surface with defined shape and dimensions, able to contain, with the required level of probability and confidence, any hazardous debris produced by the fall of a suborbital vehicle or its parts during the flight phases included in the suborbital operations. Sezione definizioni del regolamento.

[9] V. Infra par. 5.1. Responsibility for the activities.

[10] Section B, Chap. 7, par. 6.3, Regulation on Construction and Operations of Spaceports.

[11] The document defines "risk" the possibility that a specific event may occur and the related consequence, i.e. the possibility of any loss or damage, measured in terms of severity and probability.

[12] Section B, Chap. 7, par. 1, Regulation on Construction and Operations of Spaceports.

certificate. It is also, of course, an essential prerequisite for emergency management planning and it is implemented by the Emergency Plan.[13]

Such analysis is elaborated from the spaceport operator,[14] who collaborates with the public subjects and the operators interested and/or directly involved in the activities of pre-flight, flight and post suborbital flight. The aim is, of course, to control the dangers generated during operations at the spaceport in order to safeguard human lives, property and the environment inside, in the areas surrounding the spaceport and in the associated flight corridors, and ensure the continuity of the operation of the spaceport and the airport.

Special attention is given, in the Space Port Emergency Plan, to the prevention of the specific risk of fires and explosions.[15] The spaceport must be designed, implemented and managed in order to:

(a) minimise the causes of fire and explosion;
(b) limit fire propagation and the explosion effects;
(c) ensure the possibility that the persons will leave unscathed the dangerous area, or interested area by the event, or that they may be rescued in any other way;
(d) ensure the possibility to rescue teams to operate in safe conditions.

Then there are the related rescue and firefighting operations,[16] as well as the monitoring of areas adjacent to the spaceport.[17]

4.3 *Security*

Another important aim in the management of spaceport is to protect it from wrongdoing.

In particular, the operator has to assess the impact, in terms of security, of the suborbital operations that take place at the spaceport, taking into account, of course, the airport structure where it is located. It must:

> to prevent the non-authorised access to flight infrastructures, installations subject to danger of explosion or fire and any other area of the spaceport that, otherwise, is not controlled by the suborbital vehicle operator (on the basis of the agreement with the spaceport operator), through security personnel, surveillance systems, physical barriers or other means approved by ENAC during the certification process of the spaceport.[18]

---

[13] For deepening v. Section B, Chap. 7, par. 4, Regulation on Construction and Operations of Spaceports.

[14] V. Infra par 5.1 Responsibility for the activities.

[15] Section B, Chap. 7, par. 3, Regulation on Construction and Operations of Spaceports.

[16] Section B, Chap. 7, par. 5, Regulation on Construction and Operations of Spaceports.

[17] Section B, Chap. 7, par. 7, Regulation on Construction and Operations of Spaceports.

[18] Section B, Chap. 8, par. 6, Regulation on Construction and Operations of Spaceports.

In addition, he elaborates and implements the "Spaceport Security Program", which integrates the contents of the Airport Security Program where he operates, regarding the peculiar aspects of suborbital operations.

With regard to the entry of mail, cargo, supplies, operators and equipment of the operator or external and for flight staff, the spaceport is subjected to the same international and national rules as those applicable to airports open to commercial air traffic; for the occupants of a sub-orbital flight and for participants in any capacity other than the crew, it is applicable the same international and national rules foreseen for passenger security checks at airports open to commercial traffic, with stations dedicated to the control of people and hand luggage.[19]

5 The Most Delicate Legal Details Related to the Establishment of Such Structures

One of the most sensitive aspects in the management of high-risk activities is responsibility.

The question is related to the necessary obtaining, by the person identified as responsible, of a certificate which presupposes the existence of the structural and management requirements necessary to allow proper management of the plant and safe operation of the activities.

5.1 Responsibility for the Activities

The reference entity in terms of responsibility is identified by the Regulation in the Spaceport Operator:

> The legal entity that, together with other activities or exclusively, is entrusted with the task of administer and manage the spaceport infrastructures, as well as coordinate and control the activities of the different vehicle operators that operates within the and infrastructures, including the state property it got entrusted with, and from which it takes the related responsibilities. The spaceport operator is also the subject entitled to apply to ENAC in order to obtain the spaceport certification.[20]

The operator shall be the same operator of the airport where the spaceport is located.

This entity is primarily responsible for safety within the infrastructure as well as during the operations of the suborbital vehicle; in brief, for the safe operation and maintenance of the spaceport.[21]

[19] Section B, Chap. 8, Regulation on Construction and Operations of Spaceports.

[20] Section A, Definitions, page 9, Regulation on Construction and Operations of Spaceports.

[21] Section B, Chap. 3, par. 3, Regulation on Construction and Operations of Spaceports.

He shall also ensure that the operator of the suborbital vehicle has the necessary authorisations; in particular, he has to coordinate with the provider of air navigation services related to suborbital operations.[22]

In case of emergency[23] it must operate in synergy with the responsible subjects and/or involved in the intervention of rescue and firefighting, for the civil protection and the rescue at sea, as established in the emergency plan.[24]

For matters not covered in the regulation, it recalls the obligations and responsibilities of the operator contained in Regulation (EU) No 139/2014, Annex III, Chap. C (ADR.OR.C) and in the corresponding AMC and GM of EASA ED Decision 2014/012/R, applied to spaceport.[25]

5.2 *The Spaceport Certification*

The spaceport certificate[26] attests the compliance of the infrastructures and the management system of the spaceport with the requirements of the Regulation and, in particular, certifies that:

(a) the company organisation, the means, the personnel, the management procedures and the other necessary elements for a correct spaceport management and safety are adequate to execute suborbital operations;
(b) the physical characteristics of the spaceport (infrastructures and systems), surrounding areas and associated flight corridors allow the execution of the reference suborbital operations with the reference suborbital vehicle, as provided for in this regulation;
(c) the Spaceport Manual is compliant with the provisions included in the previous Chap. 4.[27]

[22] Section B, Chap. 3, par. 4, Regulation on Construction and Operations of Spaceports.

[23] Section B, Chap. 3, par. 6, Regulation on Construction and Operations of Spaceports.

[24] V. Supra par 4.3, Security.

[25] Section B, Chap. 3, par. 9, Regulation on Construction and Operations of Spaceports.

[26] The Certification Specification includes:

(a) the identification of the reference suborbital vehicle, reference suborbital operation and highest number of suborbital operations per year per typology of suborbital vehicle;
(b) the aerodrome ICAO code which is applicable to the spaceport;
(c) the main technical and operational characteristics of the spaceport;
(d) the service providers for ATS, apron and fire service management;
(e) references regarding the auxiliary electrical supply of the visual aids for navigation in the service of the spaceport;
(f) operational conditions and limitations;
(g) the Firefighting category;

[27] Section B, Chap. 9, par. 1.1, Regulation on Construction and Operations of Spaceports.

The certificate refers to the spaceport operator as the holder of the concession to manage the spaceport (and, as we have seen in the previous paragraph) already managing the airport where the spaceport operates.

In order to obtain the certificate, the applicant shall submit an application in the form and manner established by ENAC, providing evidence that he meets all the regulatory requirements (technical and operational characteristics of the spaceport, its organisation and management system). Upon receipt of the application, ENAC examines the submitted documentation and verifies compliance with the applicable requirements.[28]

---

[28] The certification is issued by ENAC to the applicant spaceport operator on the basis of the favourable results of its verifications when:

(a) the site is designated as spaceport by the Minister of Infrastructure and Transport;
(b) the applicant for the certification of spaceport owns a certification of aerodrome where the spaceport is located, issued under regulation (EU) no. 139/2014 for aerodrome code adequate for the reference suborbital operations and vehicle;
(c) the correspondence of the Basis of certification with the requirements of the Specification of Certification established in this Regulation (see Chap. 2) is verified. It is included the compliance verification for the potential cases of Equivalent Level of Safety (ELoS), Special Conditions (CS) and existing Deviations from certification specifications (DAAD) authorised for the aerodrome in the context of the related certification or provided for in the application for spaceport certification;
(d) the applicant proved the compliances of the technical and physical characteristics of the spaceport with the notified Basis of certification;
(e) the applicant for the certification proved the compliances of its organisation and management system with the requirements included in this Regulation;
(f) applicant for the certification has:
(g) identified at least one reference suborbital vehicle and one reference suborbital operation;
(h) proved that the reference suborbital vehicle can safely operate in the spaceport and that the reference suborbital operation can be safely executed;
(i) identified the flight corridors suitable to execute the flight of the reference suborbital vehicle as part of the reference suborbital operation;
(j) proved that the risk for Third Parties on the ground within the spaceport and in the associated flight corridors is not exceeding the one prescribed by the requirements of this Regulation in terms of collective risk per mission, individual risk per year and risk for the critical infrastructures in the flight corridor, taking into consideration the potential ELOS considered acceptable by ENAC;
(k) the applicant for the certification obtained the advices and clearance by the competent National Department of Fire Brigades regarding fire prevention and protection for infrastructures, systems and activities connected with spaceport management.

The spaceport certificate is valid for an unlimited period under specific conditions.[29] ENAC may decide to limit, suspend or revoke the certificate in the event of non-compliance with the commitments and the requirements of the regulation.[30]

6 Conclusion

Italy with the birth of its first spaceport, catapults itself towards a future already present.

The Regulation is the first step towards a regulation that will contribute to the creation of the Italian regulatory framework for sub-orbital commercial transport. One of the primary aims is evidently to enrich the national infrastructure offering aimed at the development of the entire aviation system. But there is more than that. This structure is the result of the strategic design pursued by a State, which chooses to focus on innovation, research, and progress. It is the perfect meeting of economic, industrial and technological interests.

A fast track to outer space will allow the channelling of know-how and fruitful international collaborations; it is an unprecedented boost to the space economy. It is an extraordinary instrument that will facilitate space operations, with a positive impact on the territory (not only national, but also European), in terms of employment and specialization of personnel.

Certainly, this project represents a great achievement, with extraordinary potential, but also an important challenge in economic and management terms.

In particular, coordination with territorial bodies involved in the various operations, may be an obstacle to the orderly management of the spaceport, if not effectively

[29] (a) the spaceport operator maintains the compliance with the pertinent requirements of this Regulation and, in particular, the spaceport maintains the compliance with the Certification Basis, taking into consideration the provisions related to handling of findings (as established in the regulation (EU) no. 139/2014 - ADR.OR.C.020 and related AMC – GM);

(b) ENAC or other organisations entrusted by ENAC are granted access, in occasion of ordinary and extraordinary inspections, to the infrastructures, spaceport operator organisation and organisation of the external subjects that carry out activities on behalf of the spaceport operator.

This to verify the maintenance of the requirements compliance of this regulation (as established in the regulation (EU) no. 139/2014 - ADR.OR.C.015 and related AMC – GM) or to control and supervise any activities carried out by anyone and related to the operation of the spaceport and the activities that take place in it, including those of the suborbital vehicle operator;

(c) events that can affect the regulatory safety and security conditions do not take place;

(d) the spaceport operator notifies its intention to transfer the spaceport management to a different subject.

[30] For the purposes of issuing and maintaining the spaceport certification, ENAC operates following the provisions for aerodromes included in the Attachment II to the regulation (EU) no.139/2014 and the correspondent AMC and GM from ED Decision 2014/012/R EASA, applied to spaceport.

implemented (for example, the management of any disasters caused by the activities carried out in the structure, which require a high degree of cooperation among multiple actors, in order to remove or limit their effects[31]).

In conclusion, Spaceport is at the same time an important milestone and a starting point for an exciting new phase in the aerospace industry.

Maura Zara is a Space Law Researcher at First Steps Legal. She is currently graduating her Law Masters at the Università degli studi di Cagliari, Italy. She is deeply passionate about space law with a keen interest in intellectual property, protection of privacy, international contracts and arbitration. She went to China to pursue her research on International Disputes Resolution in space law. As a visiting student at Beijing Institute of Technology she also cooperated with a research group for legal framework for the APSCO (Asia-Pacific Space Cooperation Organisation) focused on the legal issues related to sharing data and the protection of intellectual property.

[31] V. supra, par. 4.2. The prevention and management of the risks.

Air Traffic and Space Traffic Management Perspective on Spaceport Operations

Alper Oren

Abstract The unprecedented pace of emerging and disruptive technologies has continued to open new horizons in the field of aviation and space, as in every field. Although it is well-known to human beings that space is an accessible domain, Virgin Galactic's and Blue Origin's flights -were conducted in 11th and 20th of July 2021, respectively- not only have shown but also have proven that space is actually closer than we think. It will not be surprising the integration of the commercial space transportation method to the next generation transportation means, considering the development and accessibility of aviation from its infancy to maturity. The development of the aviation and space from infancy to maturity has faced many opportunities and challenges during the decades, but the aviation has made itself accepted among the other transportation systems. However, the commercial space transportation is still in its infancy period and can be said that poses various challenges. Ensuring the safe and secure accommodation and integration of the suborbital flights should be considered as a hot topic that will come up frequently in the near and medium term. This chapter discusses the challenge of these emergent operations from air and space traffic management perspective.

1 Introduction

The activities in both the aviation and the space domains have grown more rapidly than most other domains through the decades. Having considered the steady growth of aviation and space related activities, the current and future requirements of those domain have been reviewed regularly and methodically. There are many different milestones in the aviation and space history. However, some of them created a significant change on the records and can be said that opened a new age. The appearance of new actors and new concepts in the broaden aviation and space activities around the world have led the creation of new opportunities and challenges. Not only aviation both also space stakeholders work on to find appropriate way and means to

A. Oren (✉)
Faculty of Aeronautics and Astronautics, Eskisehir Technical University, 26555 Eskisehir, Turkey
e-mail: alperoren@eskisehir.edu.tr

A. Froehlich (ed.), *Spaceports in Europe*, Studies in Space Policy 34,
https://doi.org/10.1007/978-3-030-88311-9_8

cope with the "new"s, namely entrants, actors, concepts and challenges. Therefore, existing standards, rules and procedures and rules can lead the accommodation of integration of the "new"s.

The first steps of aviation were taken with military aviation activities in the early 1910s. From the beginning of World War I to the end of the World War II, the rapid advancement in aircraft technology triggered the incremental improvements in aviation.

Since the beginning of aviation to the present day, except for the global crises such as 9/11, Eyjafjallajokull, COVID-19 etc., there has been a steady growth in aviation activities. The introduction of turbo-propeller aircraft could be considered as the starting point for the outbreak of diversify. This was followed by transatlantic jets in 1958, wide-bodied aircraft and high by-pass engines in the 1970s. The last quarter of the twentieth century witnessed the spreading of the digital alteration not only in-flight avionics both also in aircraft design and manufacturing techniques. The twenty-first century has opened a new horizon to the aviation, which will be discussed later of this article.

There has been a distinct transition on airspace operation from lower to higher density as a result of the transformation of technology and the growth in aviation activities. While the demand has been increasing steadily for airspace utilization, this has brought the accommodation and capacity concerns. The balance between air traffic demand and airspace capacity should be achieved to ensure safe, secure, efficient and environmentally sustainable aviation activities. Therefore, the main objective is to ensure and maintain to safe, orderly, and expeditious air traffic operations while minimizing delays.

The perpetual growth of aviation activities has resulted in congested and complex airspace. This congestion and complexity have brought concerns about both the airspace safety and efficiency, and optimal utilization of airspace. The International Civil Aviation Organization (ICAO) is responsible to assist in organizing the standards, rules and procedures from the strategic level to the operational level including the technical aspect of aviation in a global context. Its efforts are supported by regional and national organization.

ICAO has described the "Air Traffic Management (ATM)" as *"The dynamic, integrated management of air traffic and airspace including air traffic services, airspace management and air traffic flow management—safely, economically and efficiently—through the provision of facilities and seamless services in collaboration with all parties and involving airborne and ground-based functions."* which encompasses the air traffic services (ATS), the airspace management (ASM), and the air traffic flow and capacity management (ATFCM).[1]

The essential services for conducting/performing operations in airspace are provided by ATM and the essence of ATM is to ensure and maintain the safe and

[1] International Civil Aviation Organization, Doc 4444 Procedures for Air Navigation Services- Air Traffic Management, Sixth Edition, Montreal, 2016.

orderly execution of operations. On the other hand, ATM is also responsible to find appropriate balance between the growing demand and constant airspace volume while considering environmental impact and cost-efficiency.

ICAO's vision for the global ATM operational concept is integrated, harmonized and globally interoperable ATM system. The reliable ATM system is based on the provision of globally integrated services. Currently, the ATM operational concept encompasses seven components respectively, "airspace organization and management", "aerodrome operations", "demand and capacity balancing", "traffic synchronization", "airspace user operations", "conflict management" and "ATM service delivery management".[2]

On the other hand, each member of the ATM community has special roles and responsibilities to ensure the security, safety, sustainability, and stability of airspace operation. This community includes but not limited to aerodrome community, airspace providers, airspace users, ATM service providers, ATM support industry, regulatory authorities, and states.

On the contrary of air traffic management, intensive discussion on defining of space traffic management have been still on-going during the years yet have not been successful. Even though not globally agreed, The International Academy of Astronautics' (IAA) generally accepted definition of Space Traffic Management is *"the set of technical and regulatory provisions for promoting safe access into outer space, operations in outer space and return from outer space to Earth free from physical or radio-frequency interference."*[3]

Space traffic management should focus on to support safe and secure access to space for all stakeholders. This derives from the universal freedom to use outer space as stipulated in the OST. Therefore, any standards, rules and procedures related to STM should be clear and prevent the any potential misinterpretation. Although the precise and sharp distinction between aviation and space activities has not been achieved, there may have the possibility to create an analogy in terms of operational and technical aspects.

2 Higher Airspace Lower Space and New Space Actors

The commercial actors' involvement to the space activities is not a new phenomenon, it can be said that this dates to the beginning of the space age. For decades, numerous commercial enterprises on a global scale have played a vital role in not only the development and expansion of space-related operations, but also have made a big contribution in the field of space-related research and development. In recent years,

[2] International Civil Aviation Organization, Doc 9854 Global Air Traffic Management Operational Concept, First Edition, Montreal, 2005.

[3] Kai-Uwe Schrogl, Corinne Jorgenson, Jana Robinson, Alexander Soucek, Space Traffic Management- Towards a Roadmap for Implementation, International Academy of Astronautics, 2018.

numerous entrepreneurs and companies seeking the opportunities to develop low-cost suborbital launch capabilities, primarily to serve the prospective commercial space tourism that are intended to provide high-quality microgravity by reaching accessing/exceeding the Karman Line.

By the advent of the new technologies in aeronautics and aerospace, aviation is enhancing its capabilities for reaching higher altitudes, flying faster and delocalising than ever before. These new technologies, together with the enhanced capabilities, are pushing the traditional boundaries of aviation. The upper level of airspace can be defined as Space, however currently there is no universally agreed definition and delimitation of Space.[4] Discussion on the issue is still under negotiation on the UN, therefore this chapter will be well beyond this discussion and focuses on the operational concept.

Considering the incremental acceleration of commercial space activities and space operations so far, both aviation and space regulatory stakeholders have started to focus on safe, secure, and efficient accommodation and integration of these operations.

A new wave of emerging and disruptive technologies began to transpire in the early 2000s, which has been characterized as "NewSpace".[5] Similarly, the ICAO has entitled the higher airspace and unmanned aircraft system (UAS) traffic management (UTM) operations as "New Entrant".[6]

The ICAOS's 13th Air Navigation conference, which was held in 2018, hosted the discussions of Operations above Flight Level 600.[7] The Secretariat provided general status about the operations above flight level 600. The term "higher airspace operations" was suggested to refer to this subject. However, some nations raised their concerns about the suggested terminology by remarking that the term of "higher" could be confusing and proposed the need of reviewing the terminology. In this context, reaching a consensus on this term could not become fact.

In this day and time, there is a diversity between the airspace structures of the States. While some States have declared their upper boundaries, others still consider unlimited. This limitation is directly related about the air traffic service provision as ICAO Chicago Convention stipulates that every State has complete and exclusive sovereignty over the airspace above its territory.[8] The ICAO Secretariat prepared a working paper in 2015 and defined suborbital flights as a flight up to a very high altitude that does not involve sending the vehicle into orbit.[9]

---

[4] Committee on the Peaceful Uses of Outer Space, A/AC.105/769/Add.1, Historical Summary on the Consideration of the Question on the Definition and Delimitation of Outer Space,, 2020.

[5] Ram S. Jakhu and Joseph N. Pelton, Global Space Governance: An International Study, Springer Publishing, 2017.

[6] International Civil Aviation Organization, Doc 10,140 Assembly Resolutions in Force, First Edition, Montreal, 2020.

[7] International Civil Aviation Organization, Doc 10,115 Thirteenth Air Navigation Conference Report, First Edition, Montreal, 2018.

[8] International Civil Aviation Organization, Doc 7300 Convention on International Civil Aviation, Ninth Edition, Montreal, 2006.

[9] International Civil Aviation Organization, 36th Session of Legal Committee, LC/36-WP/3–2, Concept of Suborbital Flights, Montreal, 2015.

Given the development on the commercial space activities and the increased demand in suborbital flights, whether air traffic management's standards, rules and procedures are to which extent applicable and feasible still remains its mystery. On the other hand, it is difficult to assume that space traffic management could be considered appropriate method to address the issue. For this reason, by taking a proactive approach and introducing a new concept regarding flight operation between air and space (independent from boundaries) can contribute to the solution.

The first layer of atmosphere, known as the troposphere, extends from 20.000 to 65.000 ft over the northern and southern poles respectively and up to approximately 48,000 ft over the equatorial regions. Above the troposphere till approximately 327.000 ft, which is known as Karman-Line, although it is possible for the contemporary aircraft engines and aircraft to operate, this requires additional measures in order to prevent efficiency lost.[10] In parallel with the increase in altitude, the density of the atmosphere also decreases. Therefore, aircraft should fly faster than orbital speed to produce aerodynamic lift.

Two new terminologies will be defined so as to ensure the integrity and continuity of this chapter: *"Higher Airspace Lower Space (HALS)"* and *"Higher Airspace Lower Space Actors (HALSAs)"* respectively. While the first of these expressions defines the operational environment, the other expresses the actors operating in/to/from this environment.

Although there is no distinct line between the air and space, airspace and space operations are conducted through the airspace and space respectively. However, new platforms such as rocket-powered vehicles operate not only in airspace but also space. While moving upward layer of atmosphere, the flight conditions and aerodynamic requirements show an alteration, and this lead the schematic and conceptual evolution of aircraft structure and propulsion systems. In view of this alteration and evolution, author suggests entitling this operational environment as *"Higher Airspace Lower Space (HALS)"*.

In other respects, it may be delimitative and mistakable to call game-changer actors as "new entrant". As "newer actors" enter the scene, the actors that we refer to as "new" will actually lose their "new" status, but that doesn't change the fact that they will continue to take their place in HALS operations. Therefore, the author would prefer to call these platforms *"HALS Actors (HALSAs)"*.

3 Higher Level Lower Space (HALS) Operational Concept

For the time being, the evolution of operations reveals that the observed trend positive, yet at a notably lower rate. Although HALS actors differ, it can be said that it is not as wide-ranging as the diversity of airspace users or orbiting objects. Main actors of HALS include but not limited to solar-powered wings, high altitudes balloon, high altitude long endurance unmanned aircraft systems, rocket-powered vehicles,

[10] Stephen Hunter, Safe Operations Above FL600, Space Traffic Management Conference, 2015.

parabolic, supersonic, and hypersonic aircraft.[11] This list does not cover the Intercontinental Ballistic Missiles (ICBMs) and Intermediate Range Ballistic Missiles (IRBMs).

Even though most operations have performed directly or under the supervision of the States governments and military authorities, proliferation and involvement of commercial stakeholders has changed the nature of the process. The current number of the HALS operations is relatively low, comparing the aviation and space operations. However, increases in HALSAs and HALS operation requires a need for consideration of many factors in order to ensure safe and secure operations.

In this point, it should be mentioned that many of HALS operations conduct/performed from and to the spaceport. Spaceports are defined as *"a ground base from which space-capable vehicles are launched either towards different orbits in outer space or to suborbital trajectories."*.[12] Although commercial space transportation still seems to be in science fiction books, it is more likely that in the coming years/decades, we will be able to make flights from one spaceport to another by supersonic/hypersonic vehicles.[13] Therefore, spaceports are "terminus a quo" and "terminus ad quem" for HALS operations. In accordance with the Center for Strategic and International Studies (CSIS), there are 22 spaceports around the world.[14] This number is relatively low when compared to airport; however, considering the frequency of HALS operation, geographic distribution and quantitative of spaceport seems adequate but possibly will expand.

As the technological innovation and emerging of new ideas continue to progress, reasonable expectation is to be able to accommodate, integrate and finally operate in a manner that will allow safe, secure and efficient.

Some of HALSAs' trajectories intersect the atmosphere or surface of the gravitating body from which it was launched, however they are unable to complete one orbital revolution.[15] The safe, secure, and efficient accommodation and integration of HALS operations into not only ATM but also STM seem to be one of the significant challenges in the first half of the twenty-first century. Even though these operations are conducted and performed at lower density today, as experience is gained, and technology evolves NAS actors will undoubtedly transition into all areas of the airspace.

---

[11] National Aeronautics and Space Administration, A Guide to Commercial Suborbital Flight Providers for Flight Opportunities, 2021.

[12] European Space Policy Institute, ESPI Briefs No.15, About the Enthusiastic Multiplication of Spaceports Worldwide, Austria, 2017.

[13] Defence News, Aerospace Startup Gets $60 M from the US Air Force to Build Hypersonic Passenger Plane, 2021, Available: https://www.defensenews.com/air/2021/08/05/aerospace-startup-gets-60m-from-the-us-air-force-to-build-hypersonic-passenger-plane/?utm_campaign=Socialflow+DFN&utm_source=Linkedin&utm_medium=social.

[14] Center for Strategic and International Studies, Spaceports of the World, USA, 2019.

[15] National Aeronautics and Space Administration Wiki, Sub-orbital Spaceflight, Available: https://nasa.fandom.com/wiki/Sub-orbital_spaceflight.

The key assumption for HALS actors is that in order to fully operate to/in/from to airspace, they comply with the existing operational standards, rules and procedures current operations while not present an undue hazard or burden to other airspace/space actors.

Therefore, it has however become increasingly clear that there is a need to develop guidance materials to address the legal, operational, and infrastructural aspects of HALS operations. While it may be possible to accommodate the current and near-term relatively lower density operations within the existing standards, rules and procedures, the tackle is resolve of the safe and secure accommodation of these flights. On the other hand, there is also requirement to set the roles and the responsibilities of the operators and service providers.

A-two step approach can be proposed in order to achieve fully safe, secure and efficient HALS operations. This approach comprises firstly accommodation then integration to meet the operational requirements. Hence, period of accommodation leads to integration and ultimately goal can be reached.

Accommodation period can be described as the period to compensate HALSAs' inability to comply within extant standards, rules, and procedures. During this period, HALS operations should be conducted/performed along with some level of adaptation or support. From regional perspective, this phase seems manageable due to the relatively low number of operations. This phase may contain normal operations, training requirements and abnormal or problem scenarios.

Integration can be referred as the period which will be ended up by initiation of fully operational phase. Therefore, this is the critical phase since HALS operations will be expected to enter the airspace structure routinely without requiring any special provisions.

The capabilities of existing standards, rules, and procedures can be used to the extent possible, to establish accommodation and integration; however, there are some objective-based requirements to be considered as listed below:

- The accommodation and integration should avoid any significant impact on current airspace operations.
- The accommodation and integration should not make concessions to aviation and space safety nor increase the risk more than an equivalent increase in aviation and space.
- The accommodation and integration should be commensurate with the potential benefit of efficiency and capacity.
- HALSAs should comply with the existing standards, rules, and procedures - to some degree- to ensuring and maintaining the operation during the accommodation and integration period.
- HALS operations should be conducted/performed by following the relevant methodology as those of aircraft while operating through the airspace.
- Service providers and rule-makers should take into consideration HALSAs additional/ supplementary requirements.

- ATM and STM strategies should balance according to the needs of HALSAs' to optimize the HALS operation without limiting the efficiency, effectiveness, and capacity.
- HALSAs should fulfil the flight characteristics and performance requirements of operating aircraft in the airspace being transited.
- Appropriate coordination and collaboration means should establish among the States, national/regional/international agencies, and organizations, as well as the industry before being fully operational.
- Additional regulatory or statutory may be imposed on the dual-use airports, if necessary.
- The appropriate stakeholders should consider the environmental impact of the HALS operation and appropriate precautions should be incorporated during the development standards, rules, and procedures.
- Secure data sharing and distribution should enable establishing a collaborative decision-making process.
- Protecting the public and imposing insurance deficiency should be addressed in order to avoid the potential damage to property.

4 Challenges for Accommodation and Integration

Aviation and space activities have originated and evolved under different circumstances. While some commonality exists across the operational perspective, the target levels of safety for each of them are not directly comparable. Historically, civilian, and military HALS operations were conducted/performed infrequently. As HALS operations are becoming more frequent, more complex, and more global in nature, States and international/regional/national agencies should start to concern about accommodation and integration. There are already some modalities about the launch and re-entry operations, HALSAs flights should consider from a different angle. HALS operation can generalize into three different categories: (1) Transit Users (2) Persistent Users and (3) Point-to-Point.[16]

While transit users transit the airspace vertically, persistent users operate with a certain amount of time throughout their allocated area, either airspace or space. On the other hand, during the Point-to-Point operations, HALSAs traverse the airspace by departing from a spaceport and arriving to a spaceport. Departure and arrival spaceport could be either the same or different.

Current HALS operations are conducted/performed within segregated airspace by relatively large volumes of airspace and enough and spare time windows. This segregation aims to protect other airspace users from potential hazards. Even as the frequency of HALS operations has increased, this approach persists due to current planning and real-time insufficiencies. As a result, contemporary methods and techniques make shortfalls to meet operational effectiveness and efficiency.

[16] Sven Kaltenhaeuser and Ruth Stilwell, The NearSpace Interface between Air and Space Traffic Management, 69 th International Astronautical Congress (IAC), Germany, 2018.

Table 4.1 The challenges for accommodation and integration

Infrastructural	Operational	Legal
• Communication • Navigation • Surveillance • Information Exchange	• Airspace Design and Allocation • Airspace Classification • Altitude Determination and Separation	• Regulatory • Licensing • Bilateral / Multilateral Agreements

Currently applied segregation approaches probably will become less feasible soon as the number of HALSAs users increases and HALS operations become congested and complex. Therefore, a strategy toward more efficient and predictable HALS operations should be provided.

HALS operations will be provided through the collaborative and cooperative integration of humans, technology, and information. This also includes the facilities and services such as communications, navigation, and surveillance system either ground or air or satellite based. Currently, there is no globally agreed way ahead however there are many efforts to tackle the challenges so as to operate seamlessly. Therefore, this section focuses on "what is current situation and in terms of ATM and STM" and "what should be envisaged".

The remainder of this section will introduce some general challenges on accommodation and integration. These challenges may be divided into three subcategories as listed Table 4.1. Each item listed in Table 4.1 will be discussed separately and a general view will be presented. Even though each challenges deserves to be separate research paper, this chapter will provide a general perspective about each of them by creating analogy from air traffic management, as well as space traffic management.

4.1 *Infrastructural*

Infrastructure systems should be seen as the key elements to ensure and maintain safe and efficient HALS operations. Although some of the existing infrastructure systems are considered suitable for the HALS operations, especially during the accommodation and integration phases; it can be said that the existing infrastructure systems may have some deficiencies to meet all requirements in terms of HALSAs characteristics.

4.1.1 Communication

From airspace perspective, while an aircraft operating in controlled airspace, two-way radio communication should be established and continued.[17] This requirement may vary depending on airspace classification. Air traffic service requirements for

[17] International Civil Aviation Organization Annex 2, Rules of Air, Tenth Edition, Montreal, 2005.

communications can be divided into four main categories: (1) Aeronautical mobile service (air-ground communications), (2) Aeronautical fixed service (ground-ground communications) (3) Surface movement control service and (4) Aeronautical radio navigation service.[18] Emerging technologies have brought not only challenges but also opportunities. ICAO has presented the Global Air Navigation Plan[19] and Global Aviation Safety Plan[20] in order both to adapt these challenges and to respond the transformation.

Similarly, effective, and efficient communication is also essential for space operations in order to ensure successful task performance and mission safety. The appropriate communication means between mission control center and space crews play a pivotal role especially during the emergency or abnormal situation that are time critical.[21] Considering the operational environment and the spacecraft's distance from earth, it is not an easy task to transmit/receive data from Earth to a spacecraft or vice versa. Currently, NASA's Space Communications and Navigation (SCaN) manages 3 communication networks namely (1) Near Earth Network (NEN), (2) Space Network (SN), and (3) Deep Space Network (DSN) across large distances.[22]

However, nature of HALS operations and HALSAs characteristics bring some challenges in term of communication. For example, when hypersonic aircraft are considered, "plasma sheath" prevents not only radio communication but also telemetry and GPS signals. This "plasma sheath" reflects the electromagnetic signals and cuts off the communication.[23] Luckily, appropriate mitigation methods such as aerodynamic shaping, magnetic window and liquid injection have been proposed.[24] On the other hand, In recent years, NASA Ames Iridium modems has been used to by NASA to mitigate communication challenges.[25]

---

[18] International Civil Aviation Organization Annex 11, Air Traffic Services, Fifteenth Edition, Montreal, 2018.

[19] International Civil Aviation Organization Doc 9750, 2016–2030 Global Air Navigation Plan, Fifth Edition, Montreal, 2016.

[20] International Civil Aviation Organization, Doc 10,004, 2020–2022 Global Aviation Safety Plan, 2020–2022 Edition, Montreal, 2019.

[21] National Aeronautics and Space Administration, Protocols for Asynchronous Communication in Space Operations: Communication Analysis, Completed Technology Project (2012–2016), Available: https://catalog.data.gov/dataset/protocols-for-asynchronous-communication-in-space-operations-communication-analysis.

[22] Ronald H. Freeman, Improving Space Operations Workshop, Journal of Space Operations & Communicator, Volume 16, Issue 4, 2019.

[23] Xiaotian Gao and Binhao Jiang, A Matching Approach to Communicate Through the Plasma Sheath Surrounding a Hypersonic Vehicle, Journal of Applied Physics Volume:117, Issue 23, 2015. https://doi.org/10.1063/1.4921751.

[24] Eric D Gillman et all, Review of Leading Approaches for Mitigating Hypersonic Vehicle Communications Blackout and a Method of Ceramic Particulate Injection Via Cathode Spot Arcs for Blackout Mitigation, National Aeronautics and Space Administration, Technical Memorandum, 2013.

[25] D.H. Atkinson et all, Advanced Suborbital Flight Communication System Based on Iridium 9523 and Zigbee Mesh Network, 2018.

4.1.2 Navigation

The growth of aviation activities increases demands on airspace and this threw the optimum utilization of available airspace into sharp relief. Development of navigation concepts such as RNAV, RNP, and PBN have been introduced to improve operational efficiency. Current navigation infrastructures can be divided into two categories which are ground-based and space-based respectively.[26] The ground-based infrastructures, which include VHF Omnidirectional Range (VOR), Distance Measuring Equipment (DME), and Non-Directional Beacon (NDB), were initially deployed to support conventional navigation methods along the Air Traffic Services (ATS) routes. However, as the density of traffic increased, these systems have failed to meet the demand. The introduction of Area Navigation (RNAV), Required Navigation Performance (RNP) and Performance Based Navigation (PBN) techniques, generally based on Global Navigation Satellite System (GNSS) positioning, have led to setting up new ATS route networks that no longer rely on ground-based infrastructures. The expansion of satellite navigation services contributes to the continued diversity of RNAV and RNP systems in different aircraft.[27] Currently, monitoring of the ground-based and space-based navigation infrastructure performance a number of tools (including DEMETER, PBN Map Tool, CNS Dashboard, AUGUR, FAS Data Block, PEGASUS) has been improved.[28]

When the issue comes to the spacecraft, it is possible to determine the trajectory, flight path, and attitude of the spacecraft by three parameters: distance, velocity, and angle respectively. Space-based infrastructure is also attractive for spacecraft navigation considering several factors such as coverage, accuracy, integrity, and price.[29]

For HALS operation, especially for supersonic and hypersonic flights, establishing the precise positioning method is still under development. Recently NASA has conducted a flight test for The Airborne Location Integrating Geospatial Navigation System (ALIGNS) which allows pilots to position their aircraft at precise distances to each other.[30]

---

[26] International Civil Aviation Organization Annex 10 Volume 1, Aeronautical Telecommunications Radio Navigations Aids, Seventh Edition, Montreal, 2018.

[27] International Civil Aviation Organization Doc 9613, Performance-Based Navigation (PBN) Manual, Fourth Edition, Montreal, 2019.

[28] EUROCONTROL, Civil-Military CNS Interoperability Roadmap Edition 3.0, Brussels, 2020.

[29] Report Concerning Space Data System Standards Navigation Data- Definitions and Conventions, Management Council of the Consultative Committee for Space Data Systems, Informational Report, Washington, 2005.

[30] National Aeronautics and Space Administration, Journal of X-Press, Volume 62 Number 12, 2020.

4.1.3 Surveillance

Surveillance is the ability to determine, track, and updating the position of aircraft the location of aircraft accurately and reliably in air traffic management. These systems process the relevant information that is received and provide knowledge of "who" is "where" and "when".[31] The surveillance systems include air-ground surveillance systems and airborne surveillance. There may be some different requirements for the provision of air traffic control considering the phases of flight. Therefore, a surveillance system's parameters are important to meet requirements. These parameters may be characterized in terms of the coverage volume, accuracy, integrity, update rate, reliability, availability.[32] It can be said that current airspace management relies on radar systems (primary radar and surveillance radar), WAM (Wide Area Multilateration) ADS-B (Automatic Dependent Surveillance-Broadcast), ADS-C (Automatic Dependent Surveillance-Contract), and TIS–B (Traffic Information Service – Broadcast).

The "Space surveillance and tracking system" is defined as *"a network of ground-based and space-based sensors capable of surveying and tracking space objects, together with processing capabilities aiming to provide data, information, and services on space objects that orbit around the Earth"*.[33] Likewise, air traffic surveillance systems, two different types of sensors, which are ground-based and space-based, have an undeniable role in tracking the space object.[34] These sensors contribute to monitoring and understanding real-time operations in the orbital environment by presenting the position of space objects and predicting the orbital paths.[35]

For HALS operation, contemporary systems can be seen as a temporary solution for accommodation and integration. Current either ground or space-based surveillance systems have some advantages, however, there are also some disadvantages considering some factors such as speed, altitude, aerodynamic structure, etc.[36]

There can be some issues to be considered when designing a surveillance system for HALS operation. Uniquely identification HALSAs should be the main target and their interoperability with current surveillance systems should be seen initial and essential factor to increase flight safety and mission effectiveness.

---

[31] International Civil Aviation Organization, Doc 9924, Aeronautical Surveillance Manual, Third Edition, Montreal, 2020.

[32] International Civil Aviation Organization, Guidance Material on Surveillance Technology Comparison, First Edition, Montreal, 2007.

[33] Regulation (EU) 2021/696 of the European Parliament and of the Council of 28 April 2021 establishing the Union Space Programme and the European Union Agency for the Space Programme and repealing Regulations (EU) No 912/2010, (EU) No 1285/2013 and (EU) No 377/2014 and Decision No 541/2014/EU.

[34] Secure World Foundation, Space Situational Awareness Fact Sheet, Washington DC, 2017.

[35] Regina Peldszus, Faucher Pascal and Kai-Uwe Schrogl, Handbook of Space Security: Policies, Applications and Programs, Springer International Publishing, 2020. https://doi.org/10.1007/978-3-030-23210-8_104.

[36] Robert Eftekari and Regulus Group Preliminary Assessment of Surveillance Alternatives for Upper Class E Traffic Management (ETM), Federal Aviation Administration Reports, 2019.

4.1.4 Information Exchange

The ATM system consists of a diverse set of applications that have been created over time for various objectives. Many specialized communication protocols, each with its own self-contained information system: onboard the aircraft, in the air traffic service unit, and so on, describe it. Many specialized communication protocols, each with its own self-contained information system: onboard the aircraft, in the air traffic service unit, and so on, describe it. Each of these interfaces is separately and locally conceived, developed, managed, and maintained at a great cost. Furthermore, most ATM systems have their own unique ways of defining, structuring, providing, and using information.[37]

The ATM system has become increasingly reliant on accurate and timely information as the growth of aviation demand, economic challenges, and environmental concerns increase. Today, System Wide Information Management (SWIM) augments communication means that includes human-to-human communication and machine-to-machine communication. This also enhances data dissemination and accessibility while also improving data quality.

The global SWIM concept aims to foster interoperability between stakeholders in a networked environment. Information services and service orientation practice, in which information services are identified, planned, defined, and executed, are the ways to attain this goal.[38]

Enhancing the practice and utility of sharing orbital data on space objects by improving the accuracy of orbital data on space objects is also one of the main concerns of State and international intergovernmental organizations. These stakeholders also encourage the development and utilization of appropriate mechanisms to enhance orbital data accuracy for space safety, over and above the adoption of common, internationally recognized standards. Relevant stakeholders seek to promote strategies and research into novel methods to increase this accuracy by recognizing that space safety strongly depends upon the accuracy of orbital and other relevant data.[39]

These strategies could include but are not limited to national and international efforts to expand the capabilities and existing and prospective sensors' geographic distribution, the employment of on-orbit tracking aids -both passive and active-, as well as the combination, integration, and validation of data from a variety of sources. Encouragement of involvement and capacity-building in this domain by developing countries with emerging space capabilities have been given great consideration during the decades.

---

[37] International Civil Aviation Organization, Doc 10,039, Manual on System Wide Information Management (SWIM) Concept, Advanced Edition, Montreal, 2015.

[38] EUROCONTROL, Specification for SWIM Information Definition, Brussels, 2017.

[39] Committee on the Peaceful Uses of Outer Space 61st Session, A/AC.105/2018/CRP.20, Guidelines for the Long-term Sustainability of Outer Space Activities of the Committee on the Peaceful Uses of Outer Space, 2018.

Although aviation and space domains are considered separately in terms of information sharing, they also have some common points to ensure flight safety. Therefore, appropriate methodologies for information sharing have been established. However, it is hard to say that they have already reached the desired level of autonomy. Besides HALS operation should integrate both aviation and space information sharing tools and means. It seems not easy at the first glance and covers some limitations. For example, information sharing tools should be designed and implemented to be globally interoperable among the airspace, space, and HALS users. Many of the present-day models for information exchange have been designed to support point-to-point or application-to-application exchanges and this limits the flexibility of inclusion of new users. Last but not least, many ATM organizations presently manage their data in silos, resulting in duplication and inconsistency.

4.2 Operational

4.2.1 Airspace Design and Allocation

Frequently, airspace design and allocation have been a compromise between all articulated objectives in order to reconcile competing requirements in airspace use between civil and military users, which leads to often too prolonged negotiations between the parties involved. As a result, objective criteria for airspace design were established so as to assure greater transparency and predictability of airspace management procedures. The nature and goal of civil and military operations are distinct.[40] Military operations are carried out for national security or defence objectives, whereas civil aviation promotes worldwide engagement and contributions to the global economy. Both sorts of operations, however, take place in the same continuum of airspace, therefore civil aviation and military stakeholders work together to assure mutual safety. Currently, the main concept of airspace management in terms of allocation is the FUA (Flexible Use of Airspace) concept. The FUA concept is based on the idea that airspace should no longer be determined as either military or civil, but rather should be treated as a single continuum that may be used flexibly on a daily basis.[41] The FUA concept makes use of airspace structures and procedures that are especially well adapted to temporary allocation and/or utilization such as Temporary Reserved Areas (TRAs), Temporary Segregated areas (TSAs) and, Conditional Routes (CDRs).

FUA concept relies on three-layer which are "Strategic", "Pre-Tactical", and "Tactical" Level respectively. The three levels correspond to civil/military ATM coordinating tasks. Each Level has a direct relationship with the others and has an impact

[40] International Civil Aviation Organization, Doc 10,088, Manual on Civil-Military Cooperation in Air Traffic Management, First Edition, Montreal, 2021.

[41] EUROCONTROL, European Route Network Improvement Plan - Part 3 Airspace Management Handbook Procedures for Airspace Management, Edition 5.8, Brussels, 2021.

on them. While "Strategic Level" consists of a national, high-level civil/military coordination body, "Pre-Tactical Level" contains of the day-to-day management and temporary allocation of airspace. Tactical Level includes the real time activation, deactivation or real time reallocation.

Besides the airspace allocation, coordination and collaboration among the stakeholders also play a critical role to ensure efficient airspace utilization. Effective coordination and collaboration are required to meet future not only airspace but also space user's requirements and demand in terms of safety, security, capacity, efficiency. This coordination and collaboration between civil and military authorities should take place at all three levels in order to improve safety and capacity, as well as efficiency.

Coordination and collaboration between civil and military authorities are not only subject to airspace management, but also space traffic management. Current space operations, especially launch/re-entry phases, are conducted/performed under specific conditions and segregated airspace techniques are in effect during these phases of space operations. Considering their frequencies, their speeds to traverse airspace, and flight profiles, segregation can be seen as a manageable and appropriate methodology.[42]

It can be said that in the near term, HALS operation will be a whole new ball game and it is surely beyond doubt that, the regional and national organizations should collaborate. HALS operations should be incorporated into routine operations and within existing procedures and infrastructure and should meet the performance criteria of airspace being traversed. Currently, HALS operations occur relatively low-density; however, for the medium-term segregation would not be a feasible solution. The aviation sector has already experienced segregation during the unmanned aircraft integration but the integration in non-segregated airspace has been set as an objective for the aviation authorities. Therefore, the lessons learned from the unmanned systems could be seen as an opportunity for HALS accommodation and integration.

4.2.2 Airspace Classification

According to the ICAO, airspace is separated into two categories: controlled and uncontrolled airspace. Controlled airspace is separated into five categories, ranging from Class A to Class E, while uncontrolled airspace is divided into two categories, Class F and Class G.

Several elements, including "Type of Flight," "Separation Provision," "Service Provision," "Speed Limitation," "Radio Communication Requirement," and "Subject to an ATC Clearance," define the requirements for flights inside each class of airspace. Airspace classification determines the applicable flying regulations and required air traffic service provisions.[43]

---

[42] Federal Aviation Administration, Commercial Space Integration into the National Airspace System Concept of Operation, Washington, 2020.

[43] International Civil Aviation Organization Annex 2, Rules of Air, Tenth Edition, Montreal, 2005.

The main purpose of adopting airspace classification methodology is to make airspace classification easier and to unify equipment and pilot requirements for IFR and VFR flights. Because of the nature of each state's unique requirements, ICAO provisions can be interpreted in a variety of ways by states to best fit their individual needs.

Once it has been determined that air traffic services will be provided in a specific portion of airspace or in airspace associated with specific aerodromes, those portions of airspace will be designated in relation to the air traffic services that will be provided.

There is a clear need for clarity in the rules governing the application of airspace classification, as this is critical to the freedom and ability of aircraft to fly in a seamless manner across different airspace classifications.

Although airspace classification is primarily intended to aid ATC in the separation of aircraft, when a large number of air safety-related incidents are recorded, a quick overall situational review is necessary, which may necessitate modifying the airspace's classification.

Today, FAA has developed a new strategy for operations that are conducted/performed above 60,000 ft mean sea level (MSL). This methodology is called "Upper-Class E Traffic Management" and aims to presents opportunities for an alternative traffic management approach to ensure safe and efficient operations.[44] Likewise, EUROCONTROL has also introduced a new concept called ECHO, European Concept for Higher airspace Operation. ECHO aims to enable safe, efficient, and scalable operations above the flight levels where conventional air traffic operates.[45]

Airspace architecture is also another important issue for HALS operations. Considering the high-density airspace, pre-established patterns contribute to the manageability of operations in terms of detection of conflict and resolution of the conflict. These pre-defined patterns differ in each phase of flight. Traditional aviation flight begins with embarkment and continues until disembarkment. During this period, phases of flight are divided into eight categories: Standing, Taxi, Takeoff, Initial Climb, En-route, Maneuvering, Approach, and Landing respectively. While Standard departure (SID) procedures are established for taking-off and initial climb, standard arrival (STAR) routes are designed for approach and landing. For the en-route phase, an ATS route structure has developed. While traditional route structures are based on ground-based navigation aids, the new PBN concept not only reduces the dependency on sensor-specific routes and procedures but also eliminates the need for ground-based infrastructure. PBN also allows efficient utilization of airspace.

However, the flight of Virgin Galactic originated new phases such as Take-Off, Release, Boost, Apogee, Re-Entry, Glide, and Landing respectively.[46] Taking into

---

[44] Federal Aviation Administration, Upper Class E Traffic Management (ETM) Concept of Operations, Washington, 2020.

[45] SESAR Joint Undertaking, European Higher Airspace Principles and Assumptions, Brussels, 2021.

[46] Virgin Galactic Official Twitter Account, Acces: https://twitter.com/virgingalactic/status/1098635524707733504.

account that some portion of Virgin Galactic's flight path penetrates the airspace, there should be harmonization of the phases. Even though HALS operations are in their initial term, airspace architecture requires to be revised and renewed.

4.2.3 Altitude Determination and Separation

Multiple airframe manufacturers have been researching and developing supersonic aircraft that will possibly cruise within the speeds range from 1.0 to 2.5 Mach at altitudes between FL550 and FL700.[47]

Taking into consideration of supersonic aircraft's mass production in the medium term, this will lead to the increment in the population of HALSAs. In air traffic control, separation is the tool and method for preventing the collision between aircraft by keeping horizontally and vertically away from each other. The main inputs for the separation are, inter alia, the 3D/4D position of the aircraft and speed. Besides the separation methodologies, vertical separation is achieved by requiring aircraft to use prescribed altimeter setting procedures.[48]

Herein, some details can be given for ease of understanding for altimeter setting. Altitude is the vertical distance that is measured from mean sea level (MSL) whereas height is measured from a specified datum. In other respects, flight level (FL) is a surface of constant atmospheric pressure, which is related to a specific pressure datum.[49] While an altimeter set to QNH indicates "altitude", QFE indicates the height above the reference datum. Likewise, a specific pressure datum setting indicates flight levels. The transition layer is the determiner of height/altitude and flight level. The transition layer is the airspace between the transition altitude and the transition level. While transition altitude expresses the vertical position of an aircraft is controlled by reference to altitudes, the utilization of the lowest flight level will be available above the transition altitude.[50]

States specify transition altitude generally between 10,000 and 20,000 ft. If any flies at or below the transition altitude, the "altitude" is the correct term to define its vertical position. Similarly, the vertical position of an aircraft that flies above the transition altitude is expressed as "flight levels".

Determination of the vertical position of aircraft is one of the key elements of separation. Besides vertical separation, there are also time-based and distance-based separation procedures are also in force.

---

[47] Robert Eftekari and Regulus Group Preliminary Assessment of Surveillance Alternatives for Upper Class E Traffic Management (ETM), Federal Aviation Administration Reports, 2019.

[48] International Civil Aviation Organization, Doc 4444 Procedures for Air Navigation Services- Air Traffic Management, Sixth Edition, Montreal, 2016.

[49] International Civil Aviation Organization, Doc 8168 Aircraft Operations Volume 1 Flight Procedures, Sixth Edition, Montreal, 2018.

[50] International Civil Aviation Organization, Doc 8168 Aircraft Operations Volume 3 Aircraft Operating Procedures, First Edition, Montreal, 2018.

However, not only the complexities but also deficiencies in communications, navigation, and surveillance systems, traditional separation methods possibly fall short. Although the separation could be seen as a tactical process for airspace operations, new separation concepts should be present at the strategic level, which should be performance and trajectory-based.

On the other hand, current altitude determination methods and also equipment/avionics are not capable to determine the HALSAs vertical position at such heights. Therefore new onboard technologies should be developed by industry. From a regulatory perspective, the relevant bodies also should take the responsibility to promote new structures for altitude determination. Currently, airspace structures consist of only one transition altitude; however, the introduction of new technologies and new altimetric methodologies could lead to the investigation of new solutions.

4.3 Legal

4.3.1 Regulatory

As commercial space launches have continued to increase since last decades, the point-to-point spaceport operation is a candidate to be the new transportation mean. Space transportation can be defined as the movement of objects or passengers, to/from space and/or spaceport. On the contrary contemporary space operation and transportation activities, space transportation is generally carried out by commercial companies. As spaceport operations develop and become more complex, so too has its regulation. In addition, a great deal of new terminology arose, frequently without commonly accepted definitions (or with contradictory ones), and with multiple terms used to the same issue. Regulation is the process of delivering an authoritative direction to achieve and maintain a desired level of direction. All regulation entails a regulatory process, which entails multiple patterns of action by individuals interacting to develop and sustain a desired result for the regulated subject or entities. Regulatory structure, i.e. the organizations or other entities engaged, and the legal framework, are also involved in all regulations (such as licenses, regulations, and agreements). Finally, all regulations have regulatory substance, which refers to the specific issues that are being governed.

4.3.2 Bilateral/Multilateral Agreements

The process and structure of space transportation regulation should have three distinct venues national, bilateral, and multilateral; therefore, appropriate interaction tools should be established among these venues. National regulation of space transportation. should regulate by a State within the compass of its territory as well as its sovereignty over that territory and the airspace above. Today, it can be said that international air transportation is regulated depending on traffic rights. It is important to express that "Five Freedom of Air" is the result of eliminating the restrictions for air transportation. These freedoms allow to *"Overfly"* (First Freedom Right), *"Technical Stop"* (Second Freedom Right), *"Set Down Traffic"* (Third Freedom Right), *"Pick Up Traffic"* (Fourth Freedom Right), *"Carry Traffic to/from Third State"* (Fifth Freedom Right) respectively. Beyond these, there are also another four freedoms. These also allow to *"Carry Traffic* via*Home State"* (Sixth Freedom Right), *"Operate from Second State to/from Third State"* (Seventh Freedom Right), *"Carry Traffic between Two Points in a Foreign State"* (Eight Freedom Right) and *"Operate Only in a Foreign State"* (Nineth Freedom Right).[51] However, ICAO characterizes last four as "so-called" since they have not been officially recognized as such by international treaty.

As the point-to-point spaceport operation aims to carry cargo, passengers, and a combination of them, these freedoms can be taken as an example for space transportation. Current space transportation flights are conducted/performed from one spaceport to the same spaceport. However, the booming of space capacities and capabilities probably will lead and trigger a boost of spaceport's number.

National space transportation regulations should also take into account the state's international duties under bilateral and multilateral agreements and arrangements. Bilateral regulation is regulation carried out jointly by two parties, most commonly two states, however, one or both parties could be a collection of states, referred to as a *supra-State.*

The completion, implementation, or continuation of some type of intergovernmental agreement or understanding governing space transportation between the territories of the two parties is usually the purpose of bilateral regulation. Formal consultation to conclude, interpret, expand, or revise an intergovernmental agreement, arrangement, or understanding involving space transportation accounts for a significant portion of intergovernmental bilateral regulatory activity.

Multilateral regulation is regulation that is carried out jointly by three or more States, either as part of an international organization and/or a multilateral treaty or accord or as a separate activity. The primary objective of multilateral regulation in spaceport transportation is the completion, execution, or continuation of common arrangements, policies, agreements, or rules on areas of mutual interest to the various parties. Communication and interaction during multinational meetings held to analyse issues, make recommendations or resolutions, or finalize or amend intergovernmental or non-governmental agreements is the most fundamental process of multilateral regulation. Ad hoc and recurring exchanges between international bodies are also part of multilateral processes.

[51] International Civil Aviation Organization, Doc 9626, Manual on the Regulation of International Air Transport, Third Edition, Montreal, 2018.

4.3.3 Licensing

ICAO has developed globally accepted standards and recommended practices on personal licensing for pilots-on-board or remote pilot, flight engineer, flight navigator, flight radiotelephone operator, aircraft maintenance (technician/engineer/mechanic), air traffic controller, flight operations officer/flight dispatcher, and aeronautical station operator.[52] The aim of licensing is to evaluate and maintain the required competency level—which is the combination of skills, knowledge, and attitudes- of the aforementioned personal in order to conduct/perform safe airspace operations. States are responsible to establish appropriate authorities that are responsible for all aspects of licensing for personal licensing. States have the right to issue their own licenses based on national regulations and convert or validate the licenses issued by other States. Today, the conversion and/or validation process in terms of issuance, maintenance of validity, and extension of privileges have been well established. Besides the personal licensing perspective, there are also some different requirements for aircraft, airfields, as well as ATM-CNS infrastructures to be met. These requirements are examined under the licensing or certification process. Either it is assumed licensing or certification, ICAO has also established common standards, and procedures and they are enhanced and adapted by the regional organizations. However, for space operations that is not the case. It is hard to there is a globally agreed licensing framework of space operations and the person who is involved in the space operations. Therefore, this issue remains a mystery. For example, the FAA has set a common guideline for space operations and released the criteria for the crew members. According to FAA, the crew is *"any employee or independent contractor of a licensee, transferee, or permittee, or of a contractor or subcontractor of a licensee, transferee, or permittee, who performs activities in the course of that employment or contract directly relating to the launch, reentry, or other operation of or in a launch vehicle or reentry vehicle that carries human beings. A crew consists of the flight crew and any remote operator."*[53] For the HALS operation, the main question would be whether the participants of those flights will be considered astronauts or not. If this is a kind of Yes/No question, then the answer should be, unfortunately, "No" in accordance with FAA's view for everyone else who flies with the HALSAs is just a "spaceflight participant" not an astronaut.

5 Conclusion

Today, HALS operations are conducted/performed in fewer numbers than airspace operations or even space operations However, as technology advances, human being interest in HALS operations increases and becomes financially accessible,

[52] International Civil Aviation Organization Annex 1, Personnel Licensing, Twelfth Edition, Montreal, 2018.

[53] Federal Aviation Administration 14 CFR Section 401.7.

the number of HALS operations will inevitably increase. This increment creates uncertainty not only for airspace users but also HALSAs. Providing safe and efficient access to airspace and HALS for both airspace users and HALSAs poses a number of challenges that may require time to be addressed.

The starting point could be the decision of HALS operations status from a regulatory perspective. It is vital to determine whether HALS operations are/will be included in airspace operations or space operations. Or should a new concept namely aerospace operations be introduced? These considerations will influence both the operational and the regulatory perspective of HALS operation.

It is inevitable that the ICAO's role on accommodation and integration of HALSAs is vital. While aviation activities are based on the mainly Chicago Convention and its annexes, space activities are governed by the five international treaties and the five sets of principles. It is no doubt that outer space activities are carried on in full conformity with said treaties and principles. However, there have been some concerns and opinions that OST cannot meet the needs and requirements contemporary space activities.

Besides the legal aspect, there are also many other challenges on the operational and infrastructural aspects. From an operational aspect, ATM and HALS traffic management processes should accommodate dynamic flight trajectories and should provide optimum operational solutions. Tt the first stage, segregation can be a feasible methodology for accommodation and integration, however, this segregation should be set in terms of the size, shape, and time so as to minimize the operational impact for extant users. Current coordination and collaboration tools and procedures are required to be enhanced to minimize any constraints on operations. The complexity of accommodation and integration may limit the degree of flexibility for the existing users, however smooth transition periods should be established for seamless operation of HALSAs.

Ground based infrastructure, including airports, should act as a single integrated structure to accommodate and integrate the growth of HALS operations and should support a better performing in an intermodal environment. Even though spaceports are the main bases for HALS operations, dual use airports are also could be considered as the main base for HALS operations. the fact remains that, dual utilization of airports is require the enhancement of infrastructure and upper structure.

Last but not least, as a starting point, HALS Traffic Management should define for the HALS operations and this definition could be as follows: "*Safe, secure and efficient accommodation and integration provision of Higher Airspace Lower Space Actors into not only the airspace but also the space considering the extant airspace and space users requirement and minimizing the operational impact but improving the interoperability throughout the airspace, space and higher airspace lower space operational environment*".

Alper Oren holds a Ph.D. in Aerospace and Aeronautics. Prior to Ph.D., he earned his M.Sc., in Electric-Electronic Engineering as well as his BSc in Electronic Engineering. He combined his knowledge of airspace management and modelling/optimization. He has military air traffic controller background. Currently, he serves at NATO HQ as military advisor on aviation related policy issues. Additionally, he holds the chairmanship of EUROCONTROL the Military Harmonisation Group (MilHaG). He is also passionate about space situational awareness and space traffic management and focuses on analysing the principles of space traffic management by creating analogy from air traffic management. He is the Fellow Member of The International Association for the Advancement of Space Safety and member of The Royal Aeronautical Society (MRAeS).

The manufacturer's authorised representative in the EU is Springer Nature Customer Service Centre GmbH, Europaplatz 3, 69115 Heidelberg, Germany. If you have any concerns regarding our products, please contact ProductSafety@springernature.com

Printed and bound by CPI Group (UK) Ltd, Croydon, CR0 4YY

30/09/2026

02225665-0005